ASTRONOMY O

*A Devastating and Complete Repudiation
of the Big Bang Fiasco*

Roy C. Martin, Jr.

University Press of America,® Inc.
Lanham • New York • Oxford

Copyright © 1999 by
University Press of America,® Inc.
4720 Boston Way
Lanham, Maryland 20706

12 Hid's Copse Rd.
Cumnor Hill, Oxford OX2 9JJ

Library of Congress Cataloging-in-Publication Data

Martin, Roy C.
Astronomy on trial : a devastating and complete repudiation of the
Big Bang Fiasco / Roy C. Martin, Jr.
p.   cm.
Includes bibliographical references.
1. Big bang theory. I. Title.
QB991.B54M37   1999 523.1'8—dc21   99—26508 CIP

ISBN 0-7618-1422-1 (pbk: alk. ppr.)

⊖™ The paper used in this publication meets the minimum
requirements of American National Standard for Information
Sciences—Permanence of Paper for Printed Library Materials,
ANSI Z39.48—1984

For Fred Hoyle and Hannes Alfvén

Quick Glance

Table of Contents

Supernova 1987A Rings

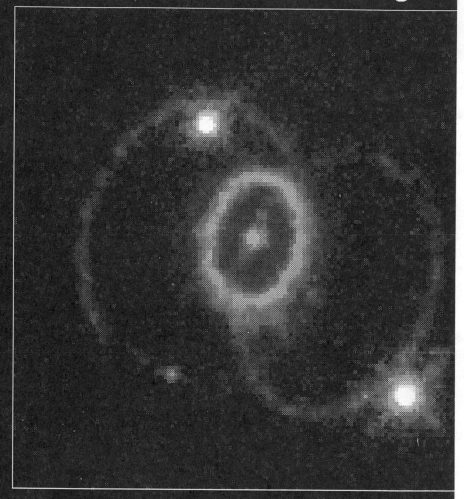

Hubble Space Telescope
Wide Field Planetary Camera 2

SPACE
TELESCOPE
SCIENCE
INSTITUTE

Preface

Supernova 1987A, which was observed in February of 1987, is a thing of beauty. The exquisite, seemingly inexplicable series of concentric rings affords such an interesting picture, that it is frequently presented as one of cosmology's spectacular discoveries. The supernova, or exploded star, which supposedly released energy equated to 100 million Suns, is a true phenomenon. Because:

> Supernovae are supposed to result in super-dense neutron stars, but none is seen.

> The ring structure is unique.

Why?

> This beautiful display is certainly noteworthy, but scientist's desperate efforts to explain the observed within a purely gravitational Big Bang framework, as reported in *Science News*, 22 February, 1997,[244] are more so.

Why is no neutron star seen? It is said:

1. Dense material that didn't get blown away hides it,
   Or,
2. The material was of just the right amount to be converted into a black hole, and they of course, can't be seen.

Why the rings? We are told:

3. The star swelled to 100 times its earlier size into a red super giant.
    Then...
4. A low-speed wind blew off the outermost layers.
    Then...
5. The star shrank into a blue super-giant, a mere 10 times the original size.
    Then...
6. The blue super-giant expelled a high-speed wind that caught up to the originally blown-off material, and when the blue wind ran into the red wind, the rings were formed.
    Then...
7. Ultraviolet radiation ionized gas in the rings, and as the gas cooled, it emitted the light seen as the rings.
    Then...
8. The rings were formed instead of shells, because the red wind was stronger in some directions than others.
    Then...
9. The uneven winds are explained by an unseen gravitational source.
    And...
10. This "simple model" explains why the remnants radiate radio waves and X-rays."

Almost all scientists, in every discipline other than cosmology, procedurally discount any complex scenario to explain anything. But this fabrication, far from simple, is instead embarrassingly contrived.

Discussion:

1. The idea that dense material could hide a neutron star that is supposed to result from the extraordinarily complete collapse of a supernova scenario is laughable.

2. "Just right" is frequently applied where the odds of such happenings are low, and everywhere else, black holes are the explanation for extraordinary disturbances and dazzling displays. Nothing like that is visible at the center of the inner ring.

4. Where did the low speed wind come from? Any burp from the star itself would tend to be spherical. If it came from without, residue would be on one side of the star only.

6. Supernova explosions are not generally referred to as mere high-speed winds: they are supposed to explode with the energy of 100 million Suns. Anything in its way wouldn't be merely ionized in a ring or any other shape, it would be completely swept away.

8. "Stronger in some directions than others" is a terribly weak explanation for the perfect symmetry displayed.

9. "Unseen gravitational influences" are used over and again, ubiquitously, as explanations for the unexplainable, throughout all of cosmology.

10. This "simple model" doesn't explain radiated radio waves and X-rays anywhere near as well as do **all** of the non-Big Bang models.

Opinion:

1. This is typical of the way cosmologists explain away the unexplainable in every single aspect of Big Bangdom, and good enough reason to take them to task.

2. Far better, clearer, more believable mechanisms for such events are available to us, and this is what this book is all about.

Conclusion:

Supernova 1987A is graphic evidence that the Big Bang scenario is bankrupt.

*Astronomy **is** on trial.*

Martin

You don't have to be a cosmologist to read this, but an understanding of basic physics would help. The use of **Appendix II Definitions Handout** might be useful, but just about every thing unfamiliar to most readers would be the very exotica of cosmology **(Chapter 13)** that this book is trying to discredit.

Prologue

Astronomy, rather, cosmology, is in trouble. It is, for the most part, beside itself. It has departed from the scientific method and its principles, and drifted into the bizarre; it has raised imaginative invention to an art form; and has shown a ready willingness to surrender or ignore fundamental laws, such as the second law of thermodynamics and the maximum speed of light, all for the apparent rationale of saving the status quo. Perhaps no "science" is receiving more self-criticism, chest-beating, and self-doubt; none other seems so lost and misdirected; trapped in debilitating dogma.

It is indeed scientists, dedicated and diligent scientists, that are engaged in cosmology, and they do the very best that they can with what they have to work with; what they are about, though, is not science so much as it is desperate pretense. Cosmology, now, is more theology than physics, or even philosophy; dogmatism is the determinant; not evidence, not logic.

Physics is the science of predicting repeatable experiences. Cosmology clearly can predict nothing whatsoever about the universe; *all* of their "predications" are merely current observations attributed to inferred past events. It certainly can't predict the Big Bang; it certainly can't repeat it. Cosmology can neither predict nor repeat black holes, pinches of spacetime, dark matter, or cosmic strings (all of which we will be discussing); their measurements are inescapably based on assumption, and their results are unavoidably mere inference and by agreement.

That this book can raise reasonable questions about so many aspects of cosmology attests to the realization that very little of today's cosmology is based on fact or even properly analyzed observation. Almost all of it is assumptions, estimations, popular agreement and unsubstantiated theory; yes, clever detective work in a lot of cases, but largely on possibly misinterpreted circumstantial evidence.

A more explicit title of this book might have been "Cosmology on Trial," but the average Earthling doesn't know what a cosmology is, and astronomy as a whole must face up to the responsibility for having let the metaphysics of cosmology overwhelm the discipline. Some of this sad situation is understandable, because of the reinforcing delusion that quantum physics can answer cosmology's conundrums.

A number of scientists within the field have said similar things in a far more qualified and substantive manner than I can muster, but they have been ignored, ridiculed, and in a number of cases actually banned from the telescope and denied funding support in retaliation for their heresy, by those that have too much to lose if cracks are allowed in the shaky foundation of modern astronomy, which is called cosmology.

As a lover of science, I want to help get cosmology back on track because our students in the schools are being misinformed, the public is being misled, and public moneys are being misspent. The arguments here are as valid as any, because they are based on the works and reports of scientists as much as are any such conclusions. Any errors of interpretation should be understandable, I think, because they are based on what is said, and a lot of what is said in cosmology today is quantum-speak, a way of talking in circles and self-supporting hopefulness.

The idea of taking astronomy to trial seems natural enough, because I feel that it is deserved. I hope that this different format might attract more attention to the discussion, and help those fighting the good fight.

Martin

Chapter 1

Rationale

---

* *Because of What They have Told Us* * *A Call For the
Scientific Method* * *Cosmology is Philosophy* *

It *is* extraordinary that someone outside of a profession could criticize
its basic and most fundamental concepts. There have been no such
efforts for chemistry, physics, geology, medicine, ... or astronomy,
why should there be, how could there be, such an effort for cosmology?

Because of what they have told us.

If cosmologists had offered explanations for their "observations"
that made sense, then no questions would have been raised. Also, if
they were, indeed, basing their findings on observations, rather than on
dogmatic, unsupported metaphysical beliefs, they wouldn't have had to
offer those nonsensical explanations.

At a 1995 New York Academy of Sciences conference, 40 scholars,
scientists, and other experts spoke to the subject *The Flight from
Science and Reason*, defended the scientific method, and urged a return
to reason and logic.[245]

This is one of those great fights that sooner or later bring about
change. Women's Lib, Gay Rights, MADD Mothers, you name it;
fired-up groups manage to initiate efforts that ultimately make a
difference. So far, ... in this effort, ... this fired-up group has only a
few members, ...but we're working on it. If books like this are able to
drum up support, perhaps that group of 40 will be successful yet.

One of the finest books on nutrition, *Fit For Life*, was written by two people who were not nutritionists. Einstein was an obscure clerk in a patent office when he wrote his Theory of Relativity, and... he said that if Maxwell had graduated college, he might never have discovered electromagnetism. The point is that someone outside can sometimes have a better perspective; when you are unencumbered with myopic vision, you can see the bigger picture.

This book argues that it is the formalized "proper-think" of cosmology that has been its downfall. My approach is admittedly non-traditional, but only an outsider can put science in a non-traditional format, and ... my scientific arguments have been attested to be valid.

Worthy of note:

> **cos.mol.o.gy**/A branch of metaphysics
> **meta.phys.ics**/A branch of philosophy
> *Webster's*

I don't know when cosmology was first defined as philosophy; it might have been so called from the very beginning; because so much of it has been untestable – and science must be testable – but I like the idea that cosmology lost it's status as a science when the Big Bang was inferred from terribly weak and flimsy evidence. There are, certainly, dedicated scientists in the field that are trying to return cosmology to the respected profession that it deserves to be. Good science is being done, but it is being ignored by "the establishment." This book, unorthodox though it is, touts good science, with good scientific and logical arguments, ... and is not apt to be criticized by the informed.

Chapter 2

We, The Jury

---

*   *Background * What We Will Discuss * How This Was Developed * History * Misled by Mathematics * It is Your Decision *

The popular model of a Big Bang concept of universe development is not only completely wrong, it is impossible.

Yes, the widely promoted Big Bang beginning for the universe as endorsed by a majority of cosmologists is not only incorrect, it is, actually, impossible.

That it could be as devotedly believed and so strongly supported has to be one of the greater mysteries of science; because after 10 years of collecting evidence, and the use of about 300 sources, a prodigious number of quotes, and the simple application of basic physics; this book: attested to by prominent scientists, makes it clear, even to the layman, that the Big Bang is preposterous. Such a statement seems unbelievable, especially since there has been almost nothing in the media that would lend the Big Bang doubt; but, sadly, amazingly, science in this case, has been in great error.

It *is* a large majority of cosmologists that support the Big Bang concept, but some of their fellow scientists vigorously contest that model, and some have made charges of even collusion and suppression of evidence in its defense. A flagrant disregard of contradictory information and reason, perhaps, has placed the majority in such embarrassment that they may be inhibited from providing us with truer findings, but the charges of collusion, suppression, and misinformation are real. There is something terribly wrong in Cosmology today, and that is what this book is about.

Here, we show:

- What the Big Bang is. **(Ch 3)**
- How the Big Bang was based on *one* bit of now proven ambiguous and unreliable evidence. **(p. 47, p.119)**
- How the Big Bang is an impossibility and violates Earthly tenants of logic and physics. **(p. 20, Ch 8)**
- How the Big Bang was to have evolved regardless of initial conditions. **(p. 19, p.112, p.132)**
- How opposite assumptions are used to justify the same result. **(p. 28, p. 63, p.65, p.67)**
- How the Big Bang is defended by the invention of imaginary and fictitious particles, entities, and forces. **(p. 59, Ch 9, Ch 13)**
- How the Big Bang's reliance on gravitational forces has not worked, and why. **(p. 76, p. 85, p. 185)**
- How cosmologists, and their reliance on certaintude and the constant correction of assumptions, have themselves, been misled. **(p. 18, p. 98)**
- How mathematical constants are used as forces. **(p.21, p. 70, p.72)**
- How the mathematics is being misinterpreted. **(p. 129, p. 132)**
- How nothing works in the Big Bang scenario, and why. **(Ch 8, p. 69)**

And, we show:

That there *are* meaningful alternative scenarios.

And, we provide:

A scenario that offers the best of the alternatives.

Hopefully, the evidence and reasoned arguments presented here will leave no doubt in your mind that the Big Bang scenario is insupportable, because it is your opinion and judgment that is being solicited.

Our arguments are based on what we have been told, in truly professional sources, news items, and popular publications. Scientists may not give much credence to sources intended for the general public, but those same scientists are quite willing to be interviewed by, and quoted in them; and the science reporters and editors are every bit as professional as are the scientists, ... and it is the lay publications that you and I, and the rest of the public, rely on for our information, ...and when we start having "wormholes" and "strings" fed to us, and a universe full of "unreal" matter, then it's time that we took them to task. Wormholes, strings, and unreal matter are just some of the exotica that is being used to justify a fabricated concept, and they are the sort of thing that we'll be discussing as we try to make this case.

You have read about the so-called Big Bang in popular magazines, you have seen it in at least three or four so-called "science" shows on television, and, it has been reported in a number of newspaper articles. We have been told over and again that the universe began as a microscopic little dot of energy, and that, suddenly, in the biggest explosion in all creation, it burst into the huge expanding cosmos that we experience today. But here, we'll be presenting information that *cosmologists* have given us that, when put in perspective, and analyzed closely, just defies reason.

We are so used to the wonders of science, that we are no longer amazed. We take frequent, constant, marvelous scientific discovery as a given. We *are* impressed, though, with the obvious genius that these scientists display. And we respect them, and feel that they are, really are, giving us the facts. Who would doubt it? We will try to show you, though, that the Big Bang concept is so full of flaws that it doesn't make any sense whatsoever, and that only desperate inventions of fancy have maintained the façade.

There *has* been a small group of scientists that has indeed tried to present these arguments in the trade publications, "through channels," as it were; but the majority of cosmologists are so myopic, dogmatic, and content with the prevailing opinion, that dissenting voices are simply ignored. Cosmologists rely on their peers for judging, and voting where research dollars go; and, if your project isn't in line with current thinking, you just don't get those research dollars, or time on the big computers or telescopes. Cosmologists, like every one else, have to satisfy the people on the funding committee or they're simply ignored. And, articles that aren't "proper-speak" just don't get published.

Cosmology is modern Astronomy, which has to be one of the very oldest of professions. Starting, perhaps, with the pyramids and Stonehenge; before they were even called astronomers, men have been studying and measuring the heavens, trying to get the answers we needed. They told the ancients when to plant their crops, defined the seasons, and gave us our calendars. They calmed our fears about comets, eclipses, and even the aurora borealis. These are the people that have correlated sunspots to global weather. They taught us how to navigate by the stars. They have given us time accurate to the billionth of a second. By studying the Sun, they taught us how fusion works – that we might have a future of limitless energy. Their ranks have included some of the greatest contributors to man and civilization, since the earliest days of record keeping: Aristotle and Galileo were some of the first to demystify the heavens; Kepler showed that the Sun was the center of our solar system and told us how to measure the motions of the planets; Newton gave us an orderly universe that lasted down through the ages.

Then, ... Einstein gave us his Theory of Relativity, and Hubble determined that the universe was expanding. From that time on, ... they weren't astronomers, they were astrophysicists, cosmologists, ... and *theorists*: people whose almost every determination is based on the rigors of exotic mathematics. ...This, then, is what this case is about. One of the key elements that I hope to develop is the idea that is *mathematics* that has led cosmologists to error!

Now, how in the world can scientists go wrong in the diligent application of mathematics? Mathematics is *the* most reliable science; it has been called the "perfect" science, and the "universal" science. The beauty – the meaning of mathematics – has provoked deep thought among philosophers since Aristotle. Mathematical truths stand alone, because they are *not* matters of fact or real existence; meaning, of course, that neither physics, nor chemistry, nor any other discipline are required to support them. It's an abstract science; of postulates and axioms that determine whether propositions are true or false based on the rules of logic. However, when applied to the "less perfect" sciences, even when the mathematics is absolutely correct, the final results can be absolutely ...incorrect!

Cosmologists use every higher level mathematical discipline that you can imagine: Higgs field equations, Relativity, string theory, quantum mechanics... all of which are beyond the average science lover; all of which are beyond me. Still, heretic that I am, I hope to show that they are indeed in error; and that it is their *mathematics* that has led them to this surely embarrassing situation.

Of course it seems very much out of the ordinary and presumptuous to criticize such a well-respected profession; but, where criticism is deserved, it must be made. We may or not be scientists, but it is we who are supposed to be served by the scientists, and not the other way around. It is we who are effected by, and have to live with, and make sense of, the results of their work. It's we who have to foot the bill.

Who better to make the case? Who better to judge it? Beliefs are not the issue here: the issue is what *your* judgment, your reasoning, your common sense dictates after you read the arguments and you get to decide whether or not we have been given the correct scenario; whether or not there was a Big Bang.

It might surprise you that even so-called non-scientific consumers might be asked to exercise judgment on such a complicated, technical question, but that is precisely the way it is supposedly to be. We do that sort of thing all the time, in just about every discipline, technical or not. If we don't like the new lighting, the new medicine, the new fabric, or automobile, we just don't buy them!   The designers, engineers, and the rest, go back to their drawing boards and start over. Why in the world – why in the universe in this case – shouldn't cosmology get the same review? This is not mere academic chitchat. Billions of dollars are spent each year in the study of the universe, and we doggone well want to get our money's worth.

The main consideration is, of course, results: results that satisfy our needs, our sense of proportion, truth. We don't always get what we want, but if we can be shown that best efforts have produced the best results attainable, we can live with that. If evidence convinces us that this is the way things are, we can handle it. But, when the unbelievable is used to defend the untenable, then we as consumers, we as voters, demand something more palatable. After this case is presented, we will be asking you, in your very best judgment, whether or not the public has been given those best efforts, those best results.

How to *conclusively* make the case that the Big Bang scenario of universe development is wrong?   Mere opinion, unsupported statements, or circumstantial evidence won't be convincing, so a decisive analysis of each aspect of the Big Bang; its birth and development, the many conflicting versions and the evidence for and against them, will be made. We will present arguments against those different renditions based on the laws of physics and logic; we will show that the evidence for the Big Bang is misinterpreted and wrong; we will show that a desperate reliance on purely imaginary inventions, such as dark matter (cold and hot), quintessence, anti-gravity, and a host of others – almost every single one of which is admittedly exotic, bizarre and unrecognized by Earthly physics and science – has become addictive; and we will show that the almost entirely mathematical derivation of the Big Bang is fundamentally insupportable.

Good science, logic, and reason preclude a Big Bang way of making universes, but there *are* quite palatable alternatives for us to consider, and we will review those.

I guess that it was Will Rogers who said that "It's not ignorance that is so bad so much as is all the things we know that ain't so."

Chapter 3

What the Hell is a Big Bang?

Edwin Hubble discovered that the seemingly more distant galaxies had a redshift that was apparently proportional to distance; that the universe looked like it was expanding!

> Redshift is like the doppler-effect: the faster a galaxy moves away, the more its light "shifts" to the red end of the spectrum – a lower frequency. Supposedly, you can tell how fast a body is moving away, and therefore how far away it is, by the amount of its redshift.

We call that redshift/distance ratio, "Hubble's Law," and, it follows that if the universe is expanding, it surely had to have started expanding from an initial point, someplace, in one great big "bang." Since there was to have been nothing before the event – no time, no space, no matter – the Big Bang idea is that the universe must have begun as a "quantum fluctuation," starting from *absolutely nothing*; a microscopic dot, really, which developed into a very hot, very dense primeval fire-ball; thought to have been hotter than 10 zillion degrees Kelvin. (We will talk about "quantum fluctuations" and the starting point a bit later.)

> Kelvin degrees are the same as Centigrade degrees, but where the Centigrade scale assigns 0° to the freezing point of water, the Kelvin scale assigns 0° to absolute zero, where there is no molecular motion what-so-ever. It's the equivalent of -273° centigrade.

Popular estimates place the big event between 15 and 20 billion years ago, and it is considered as a *pinch of spacetime*. In fact, it is usually thought of as the *beginning of time itself*. As we begin the discussion of the Big Bang, we immediately run into some complications. As it turns out, there is not just one version or scenario of Big Bang development that we have to consider, there are at least three, and variants of those! There is the *Standard* Big Bang, the *Inflationary* Model, and *Loitering* Universe. To provide you with the evidence that you need to make an informed judgment later, we will look at each.

The Standard Big Bang:

The Standard Big Bang model was based on a number of *assumptions*. That the fundamental laws of physics do not change with time; that gravity functions in accordance with the theory of relativity; that the whole thing began as an almost perfectly uniform, expanding, homogeneous, hot ball of gaseous sub-atomic particles that filled, and expanded with, all of space; a highly uniform soup of matter and energy, that was so dense, that radiation couldn't even escape until about 300,000 years later!

It was assumed, also, that the densities of matter and energy have remained almost uniform throughout the universe, and that any changes in matter or energy, as time went on, were extraordinarily smooth. And, as the scenario has it, matter has been clumping into stars and galaxies ever since, while the radiation has continued to cool – and redshift.

The Standard Big Bang model "predicted" a number of effects that seem to be quite observable. For instance:

- As the universe expanded, galaxies receded from each other with a velocity proportional to the distance between them,
- That there would be a background residue of microwave radiation, and...
- That there would be particular ratios and an abundances of certain light elements.

Since they are "observed," they are called "successes." These "observed successes" of the Standard model all relate to times later than one second after the Big Bang. The problems, however, all relate to the very first fractions of that very first second. And, ... there are *problems*!

The Standard universe develops much too fast to allow the large-scale uniformity that we observe; we have to assume that the large-scale uniformity is an initial condition. On top of that, the Standard model also requires an *opposite* assumption to explain the smaller-scale non-uniformity: the galaxies, the clumps of galaxies, and "structures" that we see. And ...the developing clumps of matter would have had to grow so quickly, what with their new-found mutual gravity and all, that only very small inhomogeneities could have existed! This is called the *smoothness* problem.

The universe-wide sameness of microwave background temperature is called the *horizon* problem, essentially, because no one can figure out how points at opposite ends of the universe could be at the same temperature. "Horizon" refers to the so-called horizon distance, which is the total distance light could have traveled since the Big Bang. The horizon problem, then, is that the horizon distance, the maximum distance light could have traveled, is a great deal less than the observable radius of the universe, ... for most of its existence!

The Standard model also has an *energy density* problem. If the energy density exceeds a certain value; the universe is said to be *closed*. If the energy density is less than the critical value, the universe is said to be *open*. And, ... if the energy density is exactly equal to that critical value, the universe is said to be *flat*. Astronomers have no idea, which of these is the true picture, and this is called the *flatness* problem.

There are others: If the Standard model is extended back to those early fractions of that first second, the model would require a number of rigid, unexplained assumptions, and it would lead to a tremendous overpopulation of magnetic monopoles. These "problems," and the very many others, are why the Inflationary model was invented.

> Monopoles are theorized isolated north or south magnetic poles; supposedly 10^{15} times as heavy as a proton. They probably do not actually exist, because they haven't yet been detected.

Another aspect worth mentioning, I guess, is that quantum theorists, or, if you prefer, particle physicists; and cosmologists, have been working much closer now-a-days in an effort to meld the Big Bang model and the Grand Unification Theory, (GUT).

> And, what is The Grand Unification Theory? It's an effort to combine three of the four basic forces thought to exist. This effort is said to be successful in explaining the asymmetry of matter and antimatter in the universe, ...but a large number of defects are created

during phase transitions, ... and, ...there are ...serious problems
associated with point-like defects ... which correspond to those
magnetic monopoles ... that I ...oh, forget it. ...Besides, ...it's not
testable.

They say – indeed, mathematics demands it – that the universe
began with equal amounts of matter and antimatter, and that they
annihilated each other, to within one in a billion, and, ... that one in a
billion that remained – the entire universe as we know it – just
happened to be matter. Antimatter can be considered to be the opposite
of matter, in that it is comprised of antiprotons (negative protons) and
positrons (positive electrons.) Now, we don't have to be
mathematicians to realize that the initial, primordial universe was one
billion times bigger in matter, *plus* one billion more in antimatter; so
there had to have been *two billion times more initial universe than what
we experience today* – all within that point the size of an electron. And,
... in the greatest act of destruction ever conceived, we annihilated
almost every single bit of it! This is just *the start* of our consideration
of mathematics' relevance to cosmology.

Now, when matter and antimatter collide, and annihilate each other,
their mass is converted to gamma-rays. (Gamma-rays are the high
energy, very short wavelength radiation that we fear in atomic
explosions.) This stupendous event would have produced one *heck* of a
lot of gamma-rays, and, indeed, that theorized effluence is credited
today with being the source of the "Microwave Background
Radiation," held out, now, to be proof of the Big Bang's validity.

The Inflationary Model:

It's the more widely accepted version, and seems to solve most of
the Standard universe problems. Remember, as was said earlier, that
the Standard model problems all have to do with the first second after
the Big Bang, and the successes all come later. Well, the only real
differences between the Standard and Inflationary models exist in that
first second! They're essentially identical after that. But that first
second seems to be receiving almost all of the attention of the
theoretical physicists. They go back to 10^{-43} seconds after the Big
Bang, which has been designated as Planck Time, in honor of Max
Planck, who first elicited quantum theory; ... and even to 10^{-45} seconds,
but no further.

The reason that supporters don't want to go back further, although they are almost "there," is because the universe would become a singularity; the temperature and density would **go to infinity**, and the laws of physics wouldn't apply! So, the façade of presenting the beginning at 10^{-45} seconds, and not zero, is all that keeps the scenario's failure from being obvious!

Earlier conditions had no effect on the Inflationary model, ... supposedly. It was at Planck Time, that gravity became a separate force; and it was about this time that, in the Inflationary model, the universe *expanded 10^{50} times*! Or more! And, ...this expansion took place almost *instantly*! The laws of physics would preclude that today, but cosmologists eagerly accepted the idea because it "worked."

Well. With the universe all "inflated;" the *horizon* distance stays well ahead of the observable universe, and *that* problem is solved. Because of the expansion, space becomes flatter; much like the surface of a balloon becomes flatter when it is inflated, and the *flatness* problem is solved. The *smoothness* problem went away when it was found that inflation smoothed out the primordial inhomogeneities, which were initially, extraordinarily small!

The Loitering Universe:

Well, where the Inflationary model is a modification of the Standard universe, Loitering is a modification of the Inflationary.[5] And, this one came about, because there is an *age* problem with both the Standard and Inflationary concepts. It seems that we are up to our ears in galaxies that are thought to be as old as 16 billion years; a bit embarrassing, as you can imagine, when you claim that the universe, itself, could be as young as 15 or (lately) even 8 billion years old.

This new proposal allows the universe to go ahead with its rapid expansion, but after about a billion years, its energy is expended, and expansion drops off almost to a stop! The universe "*loiters*!" For as much as 30 billion years!

The fortuitous loitering makes the universe older that the oldest stars, and, in addition: the loitering allows time for stars to gather into galaxies, galaxies to gather into clusters, and clusters to gather into so-called wide-ranging "structures." The loitering eventually ends, of course, and expansion as we know it continues. And, ... what causes the loitering to end and the expansion to resume? ..."Strings!"

"Cosmic strings" are said to be "one-dimensional faults in the fabric of spacetime," to use a typical bit of imaginative, but standard, vernacular.

They are hypothesized to be extremely thin, about the diameter of a proton, but of very high density and energy, and wiggle violently throughout space. Scientists think that their high gravity attracted matter into galaxies, or that their violent wiggling caused matter to condense out of empty space into sheets.

Notice:

> The Standard version proved unworkable, so the Inflationary version was presented.
> The Inflationary version proved unworkable, so the Loitering model was deemed necessary.
> The Loitering model seems preposterous, so the jury surely has room for doubt.

And, regardless of the Big Bang version chosen, there are many scenario variants: all of them are quite at odds with each other; all are hotly debated; none are given more credence than the others. Chapter 17 is dedicated to them.

Supporters refer to the "Big Bang Theory," but its very name is an example of how the language is misemployed to lend undeserved credence to the issue, Note:

1) By comparing various patterns, a physicist may be led to a tentative conclusion, called an *hypothesis*, about the nature of a physical system.
2) A set of confirmed hypotheses all concerning the same system, may be combined to develop a *theory* for the behavior of that system.
3) The results of experiments may be stated as physical *laws*.

Yet there is *nothing* confirmed about the Big Bang. It is not a theory, and only weakly might it be referred to as an hypothesis. And, the physical *laws* of conservation of matter, inertia, and others that are inviolable here on Earth, are negated in space by ... mathematics. And, there are a number of non-Big Bang versions that have to be looked at also: the *Steady State* Model, and the *Quasi-Steady State* version, ... among others!

And, ...there are others...

> "...physicists remain uncomfortable with several aspects of the Inflation scenario."[17]

Chapter 4

The Situation

---

* A story * Think About It * Assumptions, by Type of Scenario * Measurement Techniques *

Let's begin with a short story:

* * *

On an exploratory mission from planet Biosphere, the space ship *End-a-prize approached the third planet from the star.*

"Position, Mr. Sooloo?"

"150 million keelomarks from the star, and level at 3 keelomarks above the planet's surface, Captain Kkirk."

"Observations, Mr. Sooloo?"

"Sir, we detected ozone at an environmental pressure of about one meelobarr and a temperature of –95 Santergrade as we descended through 80 keelomarks above the surface. The environmental pressure increased linearly as we descended, but the temperature oscillated considerably throughout."

"Odd variation, Mr. Sooloo, what do you make of that?"

"Well, Captain, perhaps there is some kind of undetectable, or 'dark' matter that absorbs energy and varies with elevation."

"Note that in the log, please."

"Of course, Captain. Here at level-off, we are experiencing 1000 meelobarrs, a temperature of 0^0 Santergade, and the composition now, ...is: 78% nitrogen, 21% oxygen, ... and the ozone is essentially zero."

"A coincidence, Mr. Sooloo? Since this environment is similar to ours on Biosphere, it may support life. Note that in the log, and Helm, Maintain a heading of 180^0 please."

"Aye, Sir."

"One-eight-zero, Sir, and I have received a report that a layer of frozen water molecules is covering the viewing ports."

"All this asymmetry and now frozen water? I confirm your dark matter analysis, Mr. Sooloo."

"Thank you, Sir, Observation: The temperature is gradually increasing on this heading and we are experiencing a drift to our left of 32 keelomarks per hour."

"A drift, you say? What do you make of that, Mr. Sooloo?"

"There appears to be some sort of attractor, or gravitational center over the horizon off to our left."

"Interesting."

'Yes, Sir, but ... the drift is suddenly measuring 161 keelomarks per hour, as if we have just entered a jet-like stream, ... and, strange as it may seem, Sir, the drift has just dropped back to about 32 again."

"That must indeed be a greatly concentrated attractor, Mr. Sooloo. Analysis confirmed."

"Thank you, Sir. I now have a report that electromagnetic sources of different frequencies and strengths are detected in all quadrants: thousand of them."

"Odd, Mr. Sooloo. Analysis?"

"Unexplained, Sir, but I doubt that such an environment could harbor life."

"We shall investigate further."

"Yes, Captain, but now ...we have lost the light from the star, and we are experiencing severe buffeting and extraordinary conditions! Our elevation is rising and falling rapidly, we are being pelted with a storm of water droplets, ...now frozen water flakes, ...and now, ...frozen water balls!"

"Maintain elevation, Helm."

"Aye, Sir."

"Now, Captain, ...we are experiencing cyclonic and tornado-like forces, ...and now ...extremely strong electromagnetic and electrostatic force fields!"

"Analysis, Mr. Sooloo?"

"Possible black hole, Captain Kkirk."

"I concur that this planet cannot harbor life, Mr. Sooloo. Take up a course to Biosphere, Helm."

"Aye, Sir."

"Captain Kkirk, as we leave, a large, swirling, aurora-like emanation is being observed over a relatively large part of this hemisphere."

"Your black hole analysis is confirmed, Mr. Sooloo."

"Thank you, Captain Kkirk."

"You are welcome, Mr. Sooloo."

"Question? Sir," asked Helm.

"Oh? Returned the surprised Kkirk.

"Might not there be natural reasons for these observations?" suggested Helm, respectfully.

Captain Kkirk did not countenance the impertinence of minority opinions that could complicate his report or cast doubt on his decisions once made, and made that obvious as he turned away and said, "Attend to your heading, Helm."

"Aye, Sir."

* * *

Granted, that this cute little story might seem silly, but it really, truly, is not. Captain Kkirk's report on his scientific expedition into space can be categorized in a manner quite similar to the report that our scientists have given us on the results of their examination of the universe; an ill-founded reliance on imaginative and exotic perception, and a formalized policy of restricting minority views.

Let us emphasize that it is not the individual scientist that we are discussing, but the profession of cosmology as a whole. Astronomy is taking its licks here, because astronomy gave birth to cosmology but failed to pass on the disciplines of the scientific method. Those 40 scholars, scientists, and other experts, that we mentioned earlier, you recall, were concerned about this very issue: the departure from the scientific method.

I read in a student's physics book that the question "Where did the Big Bang take place?" ...has no meaning. For some reason it bothered me to see for myself that our students are being misled – albeit inadvertently – even though I must have realized right along that this was so. This may seem a minor point, but it is the big picture that we are discussing here: the very biggest.

I am told that Copernicus was merely trying to improve the accuracy of his tables on planetary position when in a moment of fortuitous serendipity he moved the reference point for some astronomical functions from the Earth to the sun; however, that simple but clever move could be called the beginning of the scientific revolution, ...in spite of the fact that it took about a hundred years for its meaning to sink in. Now, you might think that the configuration of the universe is a rather "far out" concern for most of us, but changes in our understanding of the big picture brought about by great shakers like Copernicus, Darwin, and others ...Einstein, ...Hubble; shape our very thought process and our concept of who we are and where. This is no "small potatoes," ...as we say in Maine.

Let's Think About It:

Before you picked this book up, you probably had no reason what-so-ever to doubt the exciting story of a monstrously large, primeval, explosive beginning to this universe of ours, because that is just what scientists have been telling us to believe. They have told us that, over and over again. I'm guessing though, that you did not *like* the idea that this wonderful, orderly, great and beautiful universe could be the result of a terrible explosion, because nothing in our experience with, or appreciation for, nature would suggest anything of the kind. Hopefully, we can convince you that perhaps there might be reason after all to reconsider, at least, the concept of the Big Bang.

None of us will ever know for certain about the origin or configuration of the universe, but guesses or beliefs by themselves aren't enough. And unsubstantiated, obscure, exotic, unnatural inventions for the sole purpose of supporting an otherwise unjustifiable concept aren't enough either.

As we said, there are a number of different models of the universe that astronomers have postulated. Before we discuss them, shouldn't we ask ourselves: "Since there are a number of models being offered to us, and since each of these was intensively researched, justified, and proposed by one or more brilliant, experienced astrophysicists; and in spite of the talent and mathematical expertise that was used to support each of these scenarios, ...they differ. ...*Why*?"

There is no reason to speculate that the supporters of the Inflationary model were any more qualified than the supporters of the Standard version. There were, after all, a number of problems with the Standard version: Large-scale, small-scale uniformity; and the smoothness, horizon, and flatness problems, which the Inflationary model supposedly corrected. But some of those "corrections" are insufficient – at least, as far as some scientists are concerned. Certainly, those that proposed the Loitering version, ...or the "Steady State" or Quasi Steady State;" or any of the many versions that we'll talk about shortly; all question some aspects of the Inflationary model; or reject it entirely – sometimes passionately. Why?

If all of this science, mathematics, and thinking on-high results in conflicting versions of the universe, how much credence should we give to any of them? The Inflationary model is by far, the most currently popular version. But popularity is no measure. Remember, Newton's universe was "universally" popular with the world's scientists until one young clerk in a Swiss patent office, in his free time, blew them all out of the water with his Theory of Relativity.

All of the scientists believe devoutly in their particular theory. That there are different versions, though, attests that none of the theories have been provable, and all of them are questionable. Each scientist does have his favorite, just as we have our favorite football teams, but favorite teams don't always win, and favorite theories are just that.

This *is* a difficult issue, but like most complicated subjects, we can make it more manageable by taking it one bit at a time. Let's see if we can't reason together and try to determine to the best of our ability, just what this universe of ours amounts to; to *your* satisfaction. There certainly are a lot of bits to consider, so perhaps the best approach might be to begin with some of the assumptions that physicists assumed for those different models.

Assumptions:

The Standard Big Bang model, as we said, assumed a number of severe restrictions, some of which were superseded by the Inflationary model. Let's look at a few of them just a bit.

> **That Relativity Works:** Who would doubt it? But because it is so complicated, conventional methods of solving these equations work only for the most elementary cases. And, to make even these work, scientists must introduce liberal assumptions that greatly simplify Einstein's equations. So, what it amounts to, is that in support of even this most universally accepted theory, more assumptions must be made.
> **That Elementary Laws of Physics Don't Change**: Isn't this what physics is all about? Could it be suggested that the laws of physics are transient or arbitrary? And yet, this assumption was over-ruled by the Inflationary and Loitering models.
> **That Large-Scale Uniformities Must Be Assumed as an Initial Condition:** This is a recognition that scientists cannot explain why the universe is essentially the same no matter where they look, ...unless it is assumed that the beginning *was* perfectly uniform.
> **That Primordial Inhomogeneities Must Be Assumed as an Initial Condition:** This is a recognition that scientists cannot explain how matter could clump into galaxies, ...unless it is assumed that the beginning *was not* uniform.

The Inflationary model, right now the most popular, differs from the Standard model only during those first fractions of a billionth of a second, but it requires different assumptions:

Initial Conditions Can Be Arbitrary: Maybe, but as you will soon see, there is nothing arbitrary about their concept of temperatures or what was happening to the constituents before that time; and conditions during that imagined moment have been mathematically analyzed in excruciating detail.

The Grand Unification Theory (GUT): While the feasibility of the Inflationary model depends indispensably on particle theories such as the Grand Unification Theory, the theory does not include gravity, and it is not testable because it would take energies far higher than anything particle accelerators will be able to develop in the foreseeable future.

The Theory of Everything (TOE): TOE is a theoretical attempt to meld All four basic forces: electromagnetism, the atomic strong and weak forces, *and* gravity into *The* one, very most basic force. The infinitely dense, infinitely hot ultimate point is said to be the unique situation where this could have happened. The whole idea, of course, is mere conjecture, absolutely untestable, and consequently irrelevant. Many physicists felt that Einstein wasted the last 40 years of his life in search for it.[246]

The Temperature Exceeded 10^{27}K°: This is about 10^{20} times higher, or 100 billion, billion times higher, than what has been detected in stars![46] For comparison, the Standard Linear Accelerator Center (SLAC) may have exceeded 10^{13} Kelvins to find the tau particle, but that was about as high as anyone can go in such tests. So! 10^{27} Kelvins is 10,000 billion times higher than our capability to test this assumption!

The Speed of Light Can Be Exceeded in a Single Instance: It has been demonstrated over and again that the speed of light cannot be exceeded. The most powerful particle accelerators, using every bit of their capabilities, have been unable to even attain it, although they have indeed increased the mass of the particles as they approached that speed; yet we are told that physicists propose that the speed of light, fixed thereafter for the rest of eternity, could be exceeded by 10^{50} times – one time – in support of the Inflationary model. Einstein's *Relativity* would go out the window, and you'd have to use a quantum theory of gravity that *hasn't even been invented yet*. And, …it has never been explained how the mass of the entire universe, going at the fastest speed ever imagined, with all its requisite inertia, …could slow down from that high speed! Almost instantly! …This is just one of those Earthly laws of physics: inertia, that is ignored in space.

The Universe Can Start From Nothing: Thinkers well before the birth of Christ thought that this wasn't so, and almost all scientists since then, other than those specifically concerned with this issue, seem to think that this just ain't so. Another Earthly law, conservation of energy, is simply ignored, because the mathematics permits it.

The Loitering model was conceived to allow the universe enough time to form the galaxies. It assumes that after a billion years the steady expansion of the universe comes almost to a stop, for up to 30 billion years, before cosmic strings set it in motion again. Remember, now ...what it was that was supposed to allow such a happening. "Cosmic strings:" "one-dimensional faults in the fabric of spacetime," "extremely thin, about the diameter of a proton," "of very high density and energy, that wiggle violently throughout space." "Their high gravity attracted matter into galaxies." But, ...the concept of "cosmic strings" is a *mathematical construct*: a mechanism for solving mathematical problems. *It is not a force*! Even cosmologists don't consider it a force. None of them do. This is another example of how the twisting of language twists the thought-process, and once such phraseology is in common use, reality suffers, and the users tend to believe it themselves!

> "The trouble is that no current theoretical model of the evolution of the universe seems to fit all of the observations without at least some inconsistencies. Cosmologists find that they must labor to squeeze their pet theories into the steadily tightening straight jacket of observational data."[17]

The different versions certainly do differ, but there is one thread of agreement that binds them all, and that is in the assumption of an expanding universe. No, there is one other; and that is that they all do *assume* something! But, assuming anything in science is considered "iffy" at best. Clearly, whatever observations these various theories rest on can't be very conclusive. The courts don't give much credence to mere assumptions, and by the same token, we should hold any scientific theory that rests on assumptions as suspect. Assumptions are handy for leading a thought-process, but counter-productive if they lead us to the wrong conclusion, and we might not recognize that if we assume that our assumptions were correct. You and I might discuss these assumptions as we go about our analysis, but perhaps we should not give them too much weight.

Measurements:

Measurements are the toughest challenge for astronomers. Determining angle difference between bodies can seem straightforward enough, even though they can demand the most intricate of procedures sometimes, but distance measurements are by far the toughest.

All pilots know that you cannot determine the distance to a light just by its brightness. Many fighter pilots have followed a star at one time or another, thinking that it was another aircraft. Astronomers, then, must estimate distances with so called "yard-sticks" that are constructed of a combination of experience, data collection, and a clever series of inferences. For instance, by cataloging stars of a similar type, scientists are able to plot them by age, brightness, temperature; sometimes known distance, etc. into what they call sequence diagrams. Then, if you can establish the intrinsic brightness of one star on the curve, you can infer the brightness of, and perhaps the distance to, them all. Then, if you can associate one or more of these "measured" stars with groups of another, brighter type of star, similar sequencing and "stair-stepping" will permit farther and farther measurements. And, opportunities to augment these scales arise on occasion, such as that of the Crab Nebular, which was observed to explode in 1054 A.D. Knowing its age and the general velocity of its expansion, its distance can be estimated pretty accurately. Now, of course, the most-used yardstick, for the very farthest measurements is the redshift. When you meld all such efforts into a universal yardstick, you can estimate distances all the way out. It's marvelous detective work, really.

Scientists have been quite comfortable with this stair-stepping series of "yard-sticks" for a long while, long enough that they have taken it for granted. But, in spite of the fact that contrary evidence is unwelcome in cosmology, sometimes it is inescapable, such as the newly measured distance to the "Virgo cluster," here-to-fore a "crucial" standard celestial yard-stick. Recent observations with Hawaii's Mauna Kea high-resolution camera and the Hubble telescope have greatly effected measurements based on Virgo and similar references. It now seems that they might be much closer than previously thought, and distances based on them may have to be halved or something on that order. It means that the universe might be younger than the 20 or even 8 billion years suggested earlier.

So, ...the "situation" is not good! Not only do the principal versions of the Big Bang all suffer from insurmountable problems, and not only are they all based on unfounded assumptions, but the measuring mechanism for verifying any of this is admittedly unreliable.

"When you wish upon a star, will the wish come true,
if a moment later you realize ...that it was really a plane?"

                                                            Kim

Chapter 5

The Age Problem

---

* How Ages Were Determined * Anomalies * Age Problems *
When? *

When?

Favorite estimates are between 15 and 20 billion years ago, even though cosmologists would much rather support an estimation of a much earlier event. Why? Because even 20 billion years seems not enough to have allowed the observed. *Best* estimates, based on the supposed most accurate observations, place the event between 8 and 20 billion years ago, but 15 to 20 are the most often quoted times. Of course, there is no way to firmly establish such a date. To make an estimate based on just velocity, though, would demand that the cosmologist have an idea where the Big Bang took place, but he or she doesn't, so other considerations have to be factored in.

Cosmologists have cleverly stair-stepped their estimated distances to the fringes of the universe by basing their farther estimates on their closer-in estimates, but this is all tentative at best. A case in point would be the newer measurements of the distance to the Virgo cluster (that crucial yard-stick,) and as a nearer-Virgo attests, these earlier estimates could be wrong, and the universe can be even a younger *seven* billion[181] years old. Some clusters of stars right here in our own galaxy, the Milky Way, shine with the low-energy reddish light of extreme age. They are an estimated 16 billion years old, supposedly the oldest stars in the universe! Right here in our own galaxy! What a coincidence. This is a consternation, of course, if estimates are as low as 7 or 8 billion. By the same token, some of the chunks of observed material are also baffling in that it is estimated that they couldn't form in less than 20 or 30 billion years.

Are these dates merely guesses? No, determining the date of a Big Bang is far more complicated than so far mentioned. There is no consensus, and there can't be, because the poor scientists are frustrated by dramatically conflicting evidence. The Hubble Space Telescope also showed recently, that some clusters of stars that were once thought to be 10 to 15 billion years old are now less than one tenth that.

> "The universe seems too young. It's called the age problem. It seems we have structures in the universe that are older than the universe, which doesn't make sense." [5]

Can we really tell the age of an individual star by looking at it? Globular (older) star clusters are supposed to be old, but one cluster in the Milky Way shines with the uniformly bright blue light of very young stars. On the other hand, one group of blue, young looking stars is thought to be very old red giants that have had their outer layers ripped off.

Cosmologists have detected a very young star on the farthest edge of our galaxy, which was a shocker, because, supposedly, there is so little gas and dust available out there that new stars can't form. What does this say about age? Or even star formation? Another consideration, of course, might be that the young star was formed in the galactic center and was thrown there by some event! But, we'll say more about movement later.

An x-ray satellite has detected what are believed to be clusters of *quasars* (extraordinarily energetic, mysterious stars) 8 to 12 billion light-years away. In a 15 billion-year-old universe, that means they could have formed in as little as 3 billion years? If so, the universe would have started clumping-up much earlier and faster than theories can explain. Now, what does that mean? Quasars are stupendous thingamabobs – the brightest objects in the universe; emitting more light than 1000 galaxies of 100 billion stars each – supposed to contain a black hole! And clustered? Could such a spectacular object form in 3 billion years? Scientists don't think so, and theories can't explain it. Computer models indicate that 15 billion years is barely enough time to form galaxies, let alone quasars! If they *didn't* form that early, could our estimate of the universe's age be in error?

Standard theories provide a gradual transition from the uniform soup of the Big Bang to the era of galaxies, and clusters of galaxies, and the "structures" observed today. But, astronomers have detected *bunches* of distant quasars that indicate that the universe was as lumpy as it is now, when it was only 7% of its current age!

> "The cosmos has no defined beginning and could be thousands of times older than the 15 or 20 billion years indicated by the Big Bang models."[87]

In addition to these, there is a number of other interesting age-related questions for us to ponder. For instance, our Sun is estimated to be about five billion years old. It has some heavy elements in its interior, and heavy elements are supposed to be made only in the older, more massive stars. Since heavy elements are said to come only from old, more massive stars, it seems that "our" heavy elements must have come from older massive stars that went supernova. (Exploded) Whoa. How long did it take to form those earlier, old, massive stars? How long did it take for them to start manufacturing heavy elements? How long did it take for them to get to the point where they would explode into a supernova? How many supernovae did it take to supply our heavy elements, and over how long a time interval did all of the supernovae participate? How long did it take for the residue heavy elements to travel great distance through space to get here? When we've answered those questions, perhaps we can determine when our Sun started to form. And, ... does it take longer to form more massive stars? It would seem so, but are we sure? The present method of estimating star age depends crucially on our ability to calculate the evolution of stars with great accuracy, but we are talking calculations; we are talking estimates, and perhaps assumptions.

It is understood that galaxies form pretty much as a whole and some of the stars become red giants and then go supernova. This all takes a lot of time, of course, and on top of that, supernovae may not happen as often as some theories require. The more infrequent supernovae are, the longer it would take for all the heavier elements to be spread around for newer stars to pickup. To amplify the problem, all we have to do is picture the heavy elements from that exploding nova being spread out in all spherical directions. They go everywhere, delaying all the more their ultimate assimilation in newer stars. If it takes 20 billion years to make galaxies in a scenario where everything is rushing out in the same direction, how long should it take when all the raw material is spreading out every which way?

To confound the issue, clusters, and clusters of clusters that have formed the so-called "structures," had to have taken a very, very long time.

> "The large-scale structure found in the universe is ... clearly 10 times older than the 10 to 20 billion years that the Big Bang theory allows for the whole universe."[34]
> "The vast ribbons of galaxies ...are separated by nearly a billion light-years of space. So to form them, matter has to travel half that distance, if the universe was originally smooth. But galaxies are observed to travel at only about 1/600th the speed of light, so the huge structures must have taken at least 200 billion years to form."[34]

> "In its simplest form, the Big Bang scenario doesn't look like a good way to make galaxies. It allows too little time for the force of gravity by itself to gather ordinary material – neutrons, protons, and electrons – into the pattern of galaxies seen today. Yet, the theory survives for want of a better idea."[17]

So, here we range from an estimate of less than 7 billion – to at least 200 billion years ago for the big event! And, ... a model that hasn't been discussed yet: ...the Quasi Steady State universe, suggests a universe that is older than a *trillion* years, and perhaps as much as an *infinite number* of years! Odd, isn't it? Astronomers have given us a time-keeping system that is accurate to within a trillionth of a second, but the Big Bang, "the beginning of time itself, " can't be pinned down in time closer than between 7 billion, a trillion years, ... and infinity! Surely, "the beginning of time itself" has no meaning ...if we don't know when it happened.

Chapter 6

The Big Bang Scenario

---

* What Was In the Big Bang Ball?: Matter and Antimatter,
Homogeneities and Inhomogeneities, Hot and Cold Dark Matter
* The Big Bang Beginning * How Could It Happen?:
Wormholes, The quantum Fluctuation Method, the Spacetime
Method *

There is NO way to discuss the creation of universes in a physical or meaningful manner; science is not helpful in such musing, even philosophy manages to evade the subject. Cosmologists, however, insist that such an event happened, so we are left with no choice but to address the issue. Here, silliness ensues.

What Was In The Big Bang?:

We will in a moment be discussing some others of the very many totally different, hypothetical constituents and happenings that supposedly formed in that zit-size, exploding, "pinch of nothing" during that first instant, but these warrant special attention:

Matter and antimatter,
Large-scale homogeneities and small-scale inhomogeneities,
And...
Hot and cold dark Matter.

Matter and Antimatter:

Antimatter is real enough, and its possible presence in space is treated seriously, but the evidence is strong that there isn't much of it at all in the entire universe. Symmetry, and other considerations from quantum mechanics, demand that there be as much antimatter as matter; that particles, plus and minus, always exist in pairs; that they must have the same lifetimes. So, conservation laws dictate that matter *must* appear or disappear (annihilation) only as particle-antiparticle pairs. This argument is indirect evidence in support of the Big Bang because all of the observed matter "must" have had its antimatter counterpart, and if it is missing, it "must" have been used up in the Big Bang. That *anything* is left after all of that annihilation is a weakness in the argument, but, ...what the heck.

Homogeneities and Inhomogeneities:

The Standard Universe was to have developed much too fast to allow the large-scale uniformity that we observe. So we have to assume that this uniformity was an initial condition; but at the same time, a smaller scale non-uniformity that would permit the later formation of galaxies and structures of galaxies would *also* have to be a feature of that initial circumstance. Moreover, "...the developing clumps of matter grew so quickly, what with their new-found mutual gravity and all, that only very small inhomogeneities could have existed," and this is called the *smoothness* problem.

The paradox of requiring *both* extraordinary uniformity, and especially designed "very small" inhomogeneities, in the defense of the Big Bang; is ignored by using *only* one or the other, depending on which portion of it you are defending. Uniform, it should be, if everything in the universe was pressed into a pin-head-size sphere, but we need "very small ...primordial inhomogeneities," to allow the later formation of galaxies and background microwave radiation. We might ask how these assumed inhomogeneities could *survive* in a "hot, gaseous, infernal, unimaginable dense and featureless, explosive soup." Could we expect the tiny fractures in a ice cube to survive in boiling water?   ...under super-high pressure?   ...after detonating it? Mathematically, it might be possible, and when scientists justify their theories with sophisticated mathematics, we have no choice as consumers but to rely on faith; certainly, common sense is not enough in judging the physical or mathematical aspects of the universe, but still...

We were also told how the *Inflation* model subsequently solved the problem so we could live with the two, initial, seemingly contrary characteristics. It turns out that the original Inflation model had "extreme" inhomogeneous problems, but by selecting "very special" parameters, ...inhomogeneities would have been equally spread out, and ...that the inhomogeneities were homogeneous throughout!

Well. Even with "very special" parameters, this consideration, too, seems feeble. That either an homogeneity or an inhomogeneity, or both, could form in that pressure cooker, and then survive a faster than light inflation and expansion, is just one more of the many incongruities of the Big Bang defense. This is different than the impossibility of the matter-antimatter discussion, but incredulous at best. And the statement "...the Standard universe developed must too fast to allow the large-scale uniformity that we observe," can be interpreted as admission that the expansion of the universe would certainly have been explosive in nature = that there would have been a destructive rendering of any sameness, consistency, or uniformity of the universe as a whole. Not at all what we see.

Hot and Cold Dark Matter:

Along with matter, and antimatter, here, we talk about the third of this particular group of "stuff" that was to have been in the Big Bang ball: hot and cold dark matter. "Cold" has been the favored for years, but when proven unworkable, "hot" was added. This is that "universe full of unreal matter" that was mentioned earlier. Dark matter, is "hypothetical, exotic material that doesn't radiate or act with matter, except by gravity." It hasn't been proven, but Big Bang supporters desperately "need" it to explain a lot of otherwise unexplainable things, such as why galaxy arms spin as fast as they seem to do. (Or, among many other considerations, our relative movement with the Magellanics: the local group of galaxies that are our most conspicuous companions.) Cosmologists "need" dark matter to the point that they claim that there is *100 or 300 times more* of it than regular matter!

We'll have a lot more to say about this dark matter business, but for now, notice a characteristic that these three members of this particular group share: *they are all opposites*! Really, really *opposites*. The most acclaimed, most celebrated, most favored ingredients of the Big Bang ball are all opposites with each other, and absolutely unlike any of the other members of the group! And they are all to have been born together, in that tiniest of little primordial particles. In that first second or so, electrons and protons and ions were to have formed also, and in the next minute or so they would do just as much combining as conditions permitted, with precious little opportunity later as they radially separated. But how could this be? How could an electron and a proton form together? They are also opposites! Ions don't form until after protons and electrons get together, and then wouldn't all adjacent ions be of the same charge? And aren't same-charge ions very repulsive? We are told that the developing clumps of matter grew quickly with their "new-found" mutual gravity, but how could gravity start clumping matter at that first instant, without continuing to clump matter into bigger and bigger masses, continuously? And, once that first instant was over, every second, from then on, would have *decreased* the chance for *anything* to clump further.

> "Astrophysicist Simon White of the University of Arizona in Tucson, who helped develop the theory of cold dark matter, says the new survey [A sky survey by British astronomers] and other recent observations may indeed force revisions in his proposed scenario for the evolution of the universe."[22]

The Big Bang Beginning:

> "The 'beginning' refers only to our ability to describe the state of things in terms of accustomed concepts. Whether there was a creation from nothing is not a scientific question, but a matter of belief and beyond experience, as the old philosophers and theologians like Thomas Aquinas knew." [277]

The Big Bang was born *solely* because the apparent expansion of the universe surely had to have begun someplace, so a someplace was conceived. No one knows where that someplace is. The consensus has it that it was in some kind of pinch of spacetime, a spacetime defect, a fluctuation in the geometry of spacetime; *from another* universe, *...through a wormhole* (We'll discuss these momentarily) ...in that unmentionable singularity that was to have happened just before Max Planck's 10^{-43} seconds.

At 10^{-43} seconds, we are told that a *lot* must have been happening, bubbling, rapid oscillation, rapid decay, interaction, phase transitions, super-cooling and instantly re-heating, tunneling, the creation of magnetic monopoles. All physical constants: The cosmologist constant, gravitational, and such, are supposed to have had their values set in the first instant. All protons, electrons, quarks and their kin; all were defined, designed, made and duplicated, by the zintillions; all exactly alike, all obeying a set of physical constraints in accordance with an Order, ... not yet natural, ... not yet in effect.

Time, gravity, cosmic strings, phase transitions, topological defects, global textures, and the fractures that allowed the precipitation of future galaxies; all were born and survived through the yet-to-be molten stage of the Big Bang ball in its birthing. Cosmologists have decided that time began when gravity started separating from the other forces. At one moment time didn't exist, then it did. At one moment gravity didn't exist, and in the very next, it did. And, ...gravity is said to have begun its restraining effect on the apparent expansion that might cause the universe to "close" back into that original point.. someday.

It was "unimaginably hot and virtually featureless." ...I should think so, at a 100 billion times hotter than star interiors, where only the most convoluted, ingenious, imaginative use of exotic mathematics, the Higgs' fields, will describe the forever untestable situation. The Big Bang beginning is a mathematical conception that has been given credence because there is no natural physical process that can be employed. They were very "special" conditions.

Was it an explosion? If it were, it would have had to have begun as mass and converted to energy. In an explosion, the lighter particles would have been hurled further and faster than the heavier. (The law of inertia.) All the quarks, leptons, electrons would have been gone. If it began as mass, we have to explain why didn't it stay in lumps. We also have to explain how a spherically expanding gravitational source could have a decelerating effect on particles that were ejected.

> A missile launching into outer space above an Earth that retains its gravitational character, will fall back if its velocity vector is insufficient. A particle ejected from a body that spherically expands in an explosive manner, however, cannot fall back because its original velocity vector will at all times exceed the now greatly decreased gravitational pull of that widely dispersed body. This is why supernovae don't fall back together.

Even your average black hole has so much gravity that neither mass or light can get out, so how could mass get out of that original Big Bang ball which had more gravity than a zillion black holes? And, if some magical "trigger" could initiate such an event, why is that trigger not available to everyday black holes?

> "This all required fine-tuning that even scientists regard as implausible."[62]

But, supposedly it was not mass, it was all energy, because the uniform nature of the universe belies the idea that it could have been mass that exploded, and mass couldn't have existed in that extreme circumstance. Was it merely an expansion of energy? Then it would have had to begin as energy. But what mechanism, even in another universe, could have concentrated all that energy into one point? And why?

How Could It Happen?:

There is no conceivable mechanism in this universe for concentrating enough energy to satisfy the requirements of anything approaching a Big Bang event, but we are assured that it happened in another! Well, when arguments like this are presented, science and the normal thought-process are of no help.

But even "over there," how could it happen? Would some of that universe's mass and energy be gravitated into a more and more dense, increasingly massive red super-giant that would explode to form a *neutron star*, in which a piece the size of a pea would weigh a billion tons? Then, perhaps, a lot of such neutron stars would gather and make a *quasar* that would radiate more light than a thousand galaxies of 100 million Suns each? Perhaps then would collect a pile of quasars to form a *black hole*? Next, I suppose, many, many black holes and as much matter and energy as was available, would gather to be poured into that stupendous gravitational sink-hole until it was all in there, at which time, ...just as that last electron went in there, it decided to enter our universe? ..."Through a wormhole?"

It seems appropriate to apologize for such nonsensical ruminations, but how to make the arguments otherwise?

Wormholes:

As you recall...,

"They are extremely thin, about the diameter of a proton, but of very high density and energy, that wiggle violently throughout space. Scientists used to think that their high gravity attracted matter into galaxies, but I believe the current idea is that their violent wiggling causes matter to condense out of empty space into sheets."

First, a few quotes:

"You can connect one part of the universe with another, or loop around to bring a time traveler back into the past." [161]

"The Caltech machine involved travel through a wormhole, a bizarre object that physicists believe might exist at the core of a black hole." "...space would be so wrapped up that a tunnel would form, far narrower than a subatomic particle that might reach to some distant part of the universe. Anyone or anything entering the tunnel would appear instantly at the other end and, under special circumstances, would essentially travel into the past." [29]

"Most wormholes would have been attached to older universes, so our universe would have *known* what the virtual-particle density was. The system has just the right feel to it."
"Invisible, sub-microscopic rips in the fabric of spacetime."
"Other universes can interact with ours through wormholes, ...and arrange things so the cosmological constant is zero."[31]

This is the Standard thinking. Believe it or not, the above is what our scientists are telling us in scientific publications. The terms wormholes and cosmic strings tend to be used interchangeably, even though it seems that "wormhole" is meant to describe the "mouth" of the cosmic string, and "cosmic string" is more appropriately applied to "...the tunnel ...far narrower than a subatomic particle." That, indeed, is the current idea. The extraordinary concentration of energy in another universe, to be sent to ours through a wormhole, is a favored mechanism. There are billions of red super-giants, neutron stars, quasars and galaxies, and, supposedly, lots of black holes here, however, none of them seems to be coming close to forming a universe.

Whether it was mass, whether it was energy, there seems to be no meaningful mechanism for such a transfer from one universe to another, if indeed there really is more than one universe. And, if physical constraints mean anything at all in cosmology, mass or energy spewing out of a narrow "tunnel" would have been directional, much like from a garden hose, instead of the symmetrical expansion a Big Bang describes to us.

The mathematics does provide a string theory which suggests tunnel-like configurations, but surely it would be better applied to the so-called "jets" seen coming from some stars and galaxies. The mathematics is real enough, but it certainly can be misinterpreted. As an aside, one could ask, of course, "if the Big Bang started with a wormhole, why was it hot?" These questions do seem a bit silly, but there is hardly any part of the Big Bang scenario that doesn't raise such silly questions. They keep coming up, wormholes or no.

Quantum Fluctuation Method:

Let's try another approach. There is "the quantum fluctuation of nothing method." We'd have to find a place where there was no space, or matter, or time – or energy, – and in accordance with the quantum theory of gravity, ...*which hasn't been developed* yet, ...fluctuate it. Now, wait ...a ...minute. The idea of an enormous field of energy or force, possibly experiencing a fluctuation or inconsistency that permits a reflexive event, is tolerable. Our senses aren't offended by that. But it must bother us that an event could happen *without* a field of energy or force. All the energy that we have ever experienced was the result of a *happening*. Something burns, or detonates, or atomizes. Energy was released. It is very difficult for us to imagine a "virgin birth" of energy, regardless of the mathematics.

And, if that force did exist, it would have to have had far-reaching effect, just like all the forces in the universe. The forces of gravity and electromagnetic energy in space are extensive, smooth, and consistent over very long distances – billions of miles, usually – but certainly very much wider than any point. Also, the voltage or potential or gravitational differences between one point and any other within a relatively short distance are small indeed. Such potential differences couldn't light a light bulb, let alone create a universe.

If it could happen, what would touch it off? Why would that quantum fluctuate? What would be exceeded? Certainly nothing natural could demand the event. Scientists are reluctant to address such issues, because they *can't* discuss that first instant: time zero. It's that singularity bit again. It makes no sense, and they admit it! This is why the 10^{-43} seconds beginning was agreed to. Lately, though, 10^{-32} seconds is used as a starting time to make the horizon problem more workable.

Spacetime method:

Then, there is the celebrated pinch of spacetime method which appears to be the most popular among physicists; again, usually starting in another, older, larger, "parallel" universe. It seems that there are a lot of them, (One of the most natural ideas in the world, we're told.[31]) But, if that initial event was a pinch of spacetime, what would pinch it? Black holes don't seem to do it in *this* universe. What in the world does that mean? As we'll discuss later, spacetime is much more ordinary than is pretended: it's our concept of time and space that we experience in daily living. Spacetime is merely space as time passes.

No one claims that pinching space will start a universe, so "pinching spacetime" is mere verbiage. Perhaps it doesn't deserve very much of our attention, but it must be discussed, because it is held out to us as important.

Another thought: that very first, farthest particle to form on the very leading edge of the advancing wave of the rapidly expanding fire ball, is said to have indeed ...formed. The energy – earlier advancing much faster, now slowed down to almost the speed of light – is somehow given form and substance, in spite of the fact that there was now no force what-so-ever on that local bit of energy, not even acceleration. In addition, there was no Order. For any particle to form, or change in any way whatsoever, or even exist, it has to do so in accordance with natural Order. The very argument that nothing existed before the event dictates that there could not have been even natural Order. There was nothing to give order to! Does Order travel at the speed of light? Faster? The concept of natural Order trying to catch up with that advancing situation is just one more example of our forced silliness.

Well. It seems that we just don't know how we'd start the universe, but, start, it was supposed to have done.

"...there are good reasons to think that the Big Bang is seriously flawed." [6]

Chapter 7

About That Scenario

---

* Scenario Discussion * Standard Thinking * Supernatural? *
*Evidence in Support of The Big Bang: Expansion, Microwave
Background Radiation, Light Element Ratios * The beginning **

The idea that the universe was born in a single, massive explosion was first proposed in 1927 and has been supported by a majority of physicists ever since. Since the Standard and Inflationary theories agree about the Big Bang "from raison-size on up," except for the horizon distance; if we begin there, at about 10^{-32} seconds after the big event, our review can pertain to both. By the way, during that first instant, in this tiny exploding pinch of nothing, there was to have been much more than mere matter and energy: all kinds of exotica have been imagined for us to review a bit later.

Well, here, at 10^{-32} seconds, the scientists use very difficult mathematics (Higgs' fields, grand unification theories and quantum mechanics) along with some of the assumptions mentioned just a while ago, to describe the situation. Supposedly, we had just gone through some "symmetry breaking," gravity had become a distinct force, matter and antimatter began condensing, and a lot was happening.

Standard thinking has the universe experiencing an *instantaneous* inflation, a trillion, trillion, trillion, trillion times from the size of a proton to "cosmic proportions."[243] In addition to inflation, all matter and energy in the universe was to have been formed in that very first tredceillionth of a second, ...from *nothing*. It originated from a "super-cooled" state at a billion, billion, billion degrees: 100 billion times that of stars. It experienced tunneling, bubbling, oscillation, decay, false vacuums, true vacuums, and gravitational fields *that were repulsive*!

It entered its "bubble of false vacuum" state, a bizarre, extremely dense state of matter where all of its energy was stored in a mathematical concept called Higgs' fields; ...and the energy density was about 10^{60} times higher than the energy density of an atomic nucleus. ...10^{60} times greater ... is *really* greater!

This bubble of false vacuum supposedly *could develop into its own universe*, ours, or separate from ours, and as the universe expanded, more false vacuum would create itself to fill space and maintain density. Eventually the false vacuum would decay, releasing its energy and forming all of the matter in the universe in the inflationary phase.

This is the Standard thinking.

"It may indeed be possible to create a universe out of a ball of false vacuum, ...[even though] ...it's not within the range of any foreseeable technology ...sometime in the distant future, there might be some civilization that could manufacture a sphere of this form."
"It seems to be a common feature of inflationary models in that whenever you produce one universe, you end up producing an infinite number of universes."[23]

This is the Standard thinking.

We are told that it was an almost perfectly uniform, expanding, hot gas of elementary particles in thermal equilibrium – a primordial, phenomenal fireball – as protons and neutrons precipitated out of a highly uniform soup of some of the more-fundamental particles, quarks and gluons; and the behavior of matter and energy was governed by the rules of quantum physics. It would have been unimaginable dense, hot and virtually featureless, as it expanded explosively.

As the Big Bang approached 10^{-10} seconds, the four basic forces are said to have individualized, and matter and antimatter began destroying each other, not in an act of creation, but in an act of utter destruction. At about one second after the event, and for the next 10 to 15 minutes, the lightest nuclei began to form, and the radius of the universe was expanding at a now much slower 330 times the speed of light.[62] Not only was the universe expanding, *space itself* was expanding, ...they tell us. As space expands, more space pops up in between![2] (Try to picture that.) It's as if somehow more air entered a heated, expanding balloon to fill the void between molecules. The situation is popularly equated to a raison pudding on the stove. As it heats, we can picture the sauce expanding even though the raisons would not.

This also, is Standard thinking.

About one year later, the broiling gasses would have cooled to a level experienced in star interiors today, and matter and dark matter started forming. We are told that it wasn't until 300,000 years later that light began to escape from that super-density. Galaxies started making at 200 million years, (But not until density had dropped to a level far thinner then the best laboratory vacuums attainable![218]) and expansion finally settled down to just below the speed of light. From here on, based on our earlier discussion, you can pick any age for galaxy formation that you choose. Your guess seems as good as anyone's.

> A thought: if the universe began from absolutely nothing; no space, no matter, the beginning of time itself, there could have been no nature. Since it couldn't have been a ...natural act, should we call it a ...super-natural one?

Evidence in Support of the Big Bang:

There is, certainly, evidence that the Big Bang did indeed happen; evidence that supposedly can't be accounted for in competing models. Remember, the Big Bang model predicted a number of observable effects:

> That the galaxies receding from each other with a velocity proportional to the distance between them,
> That a background residue of microwave radiation exists, and...,
> There are particular ratios and abundances of certain light elements.

Let's discuss these observations just a bit.

Apparent Expansion:

Remember, redshifting is the doppler effect experienced by a light source moving away from us at high speeds. The faster a galaxy recedes, the more its light waves are shifted toward the red, or longer wavelength part of the spectrum. Because speed is related to distance in the popular expanding universe models, the higher redshifted, faster galaxies are supposed to be the farthest ones out, even perhaps to the very edge of the universe – if there is one. Sir Edwin Hubble took a small patch of sky and counted galaxies as a function of their apparent brightness. So long as the distribution of intrinsic brightness is the same in all places and at all times, he argued, the number with decreasing apparent brightness describes how redshift varies with depth. His work has been widely accepted as proof that the universe is expanding.

Redshifting, as we discussed, does suggest that galaxies do indeed move as described in an apparent expansion. Although an explosive Big Bang may not be the only mechanism for moving galaxies, just about all of the models that we have talked about so far, and even those that we'll mention later, recognize some sort of universal expansion of space. It is likened to the raison pudding.

We are asked to picture a warming, expanding raison pudding in a very large, evenly heated pot, large enough that the entire pudding can expand without interference from the sides of the pot. We can imagine the raisons separating from each other; all pretty much at the same rate, or speed. If we select one raison for our reference and observe all the others, we will note that the adjacent raisons all seem to be moving away at the standard pot rate. But the raisons twice as far away as those adjacent are in fact moving away from us, across the pot surface, at twice the pot rate, even though they are actually moving at the same speed in relation to their neighbors. Raisons five and ten times farther away than our adjacent neighbors are moving at five and ten times faster, over the pot surface, and raisons an infinite distance away are traveling at an infinite speed.   This analogy is widely used to demonstrate the "correctness" of the Big Bang scenario. Raison pudding has been the illustration pudding of choice for years.

The Microwave Background Radiation:

Predicted earlier, the discovery of the cosmic microwave background radiation that covers the sky in a rather uniform glow as a faded remainder of the grand explosion, *100 million times fainter than a birthday candle*, earned two scientists a Nobel prize. This is said to be radiation that stars or other known objects could not produce. It is now merely a weak hiss of energy of less than 3 degrees above absolute zero, or about minus 450°F, ...also predicted by theory. That background radiation was deemed important enough to launch a satellite specifically designed to study the radiation more completely: COBE, the Cosmic Background Explorer.

The radiation was one thing, but it seemed that it was too smooth to have allowed the formation of such intruders as the galaxies and structures of galaxies that are clearly here. Indeed, the observation of large "bubbles" in galactic "walls" seemed to demand a much more uneven background radiation. So, COBE was designed to be able to detect temperature variations[42] in that otherwise smooth background as small as one in 100,000. I don't know that one in 100,000 could be called "much" more uneven, but that's what they were looking for. It was felt that if COBE couldn't find those variations, then the Big Bang theory would be in trouble.

Well, at first no variation was found, but wait, COBE's accuracy wasn't too great. Finally, ...almost undetectable, ...found it was; as "primordial ripples" on the fabric of space, and "proof" of dark matter and the Big Bang: variations on the order of 30 millionths of a degree. Later, more accurate measurements were made by MAX, a balloon-launched experiment. And temperature differences of a possible 45 millionths[56] of a degree may have been detected. Scientists declared that they were looking at the beginning of time itself. The results are acclaimed as a vindication of the Big Bang scenario.

Light Element Ratios:

And then there is the abundance and ratios of light elements. Most scientists seem to believe that only a Big Bang event could have resulted in the 75% hydrogen, 25% helium mix that is detected in most of the universe today, and what is observed now matches the "predictions" of the Big Bang pretty well. Deuterium, the so-called "heavy hydrogen," is also said to be made only in a Big Bang scenario, and we do indeed have it.

Dark Matter:

In addition to the above evidence, there is the so-called lensing effect due to dark matter. We will have more to say about dark matter, but for now it is supposed to be an invisible source of gravity every bit like regular matter is, but there is, *must* be, 100 times or 300 times[143] more of the dark than the regular, to make the Big Bang scenario work. Since it is invisible, it can be detected only indirectly, but "detected" it is said to have been. More than one team has detected an arc of light, as predicted by Einstein, that apparently was caused by the bending of a distant star's beams by a nearer but invisible mass.

Antimatter:

Another bit of evidence in support of the Big Bang is the lack of a sufficient amount of detected gamma-rays: indicators of matter-antimatter collisions in the universe. (Recently, though, it's found that the Milky Way is "swimming" in them.) It is reasoned, then, that there can't be much antimatter now, in spite of quantum physics dictums that there must be equal amounts. The argument, then, is that the antimatter must have been used up in the Big Bang.

The solar wind, for example, demonstrates that all the planets in our system are made of the same stuff that makes up the Sun, because they don't radiate gamma-rays! If the planets were matter and the Sun was antimatter, the collision of solar wind particles with the planets would result in easy to detect gamma-rays.

Our most frequent visitors, cosmic ray particles, almost never indicate examples of antimatter annihilation, seeming to show (until recently) that our entire galaxy is all matter. And, since gamma-rays are not ordinarily spotted in galaxies, it is thought that they too, must be matter, although energy levels in galaxies are very close to what would be expected from antimatter annihilation. Supposedly there can't be much antimatter in the universe at all, *unless it is separated from matter by the equivalent of galaxy cluster separation.* This argument is not ordinarily made much of, in spite of the fact that it seems the most compelling.

That First Moment:

Some considerations ascribed to that very first moment were:

> "It began as a quantum fluctuation, starting from absolutely nothing."
> "A microscope dot, about the size of an electron."
> "There was no space or matter, a pinch of spacetime."
> "Going back to zero brings the universe back to infinite temperature and density in which the laws of physics don't apply."

Each of these is peculiar to say the least, but we should keep in mind that there are all kinds of strange things that are proven to exist in this world in spite of their peculiarity. In the proof, they become quite acceptable, even when they remain peculiar. We "know" that the speed of light is constant. We "know" that faith heals; that acupuncture works; that people can walk on hot coals without getting burned; that black cats shouldn't cross our path. Our sense of experience and reason has come to accept all kinds of things that we don't understand, but only after we are given proof.

But, this is why we can not feel comfortable with that first microscopic, fluctuating, bubble of a nothing; because peculiar things are not, and should not, be accepted without substantiating evidence, ...and there is no such evidence. This is that singularity that we are warned about: a meaningless mathematical aberration; an arbitrary flat, a topic of discussion, but nothing more. This infinitely tiny point of nothing, remember, was supposed to have been born in nothing – right smack dab in the middle of it!

Let's discuss the evidence.

Chapter 8

On The Evidence

Again, we noted the evidence for the Big Bang:

> That the galaxies are receding from each other with a velocity
> proportional to the distance between them,
> That a background residue of microwave radiation exists, and,
> There are particular ratios and an abundances of certain light elements.

Expansion Discussed:

The first, of course, is the observed expansion described earlier. Hubble's observations and the application of red-shifting are universally accepted as proof of expansion, and every thing seems to be racing away. Expansion, though, might be triggered by other than Big Bangs, and it seems reasonable that some parts of the universe could expand while others contract. If one part of space has a higher density or is warmer than its neighbor, then a temporary expansion would naturally be experienced. And, ...the periodic introduction of new material, such as virtual particles, is supposed to force space to expand, because Einstein told us that mass disturbs space. Even mini-bangs are a popular alternative: local explosions or sudden expansions are evidenced even in our own neighborhood!

Our local situation is noteworthy in that a super-nova explosion about 100,000 years ago, or a speculated little bang of matter and antimatter annihilation about 8 billion years ago seems to have created a hot bubble of empty space in our own back yard. (Age estimations are tough, but this is a cute coincidence.) If so, wouldn't that explain some of the expansion that we witness? Would the energy residue look like a microwave background radiation? Regardless, jumping to the conclusion that only an outlandish event could have produced the observed is not good science.

Raison Pudding Discussed:

We are asked to picture space expanding, like the pudding: but not the galaxies, like the raisons. But the analogy is weak. Raisons are relatively solid, but galaxies are relatively spacious; ...if the stars in a typical galaxy were the size of dust particles, the distance between those dust particles in a galaxy would be a number of miles! Housewives would be happy to settle for that kind of dust problem. And what is between those miles apart dust particles? Space! Yet, we are asked to picture the space surrounding galaxies as expanding, ...but not the space inside! Why not? On top of that, our local group of about 30 galaxies is over 3 million light-years across, yet it isn't expanding either, in spite of all that space in between. Why not?

Usually when this question is asked, we are told that gravitational forces within a galaxy or within a cluster of galaxies restrict the expansion of space within. But the miles-apart dust bits that we can envision here on Earth would be carried away with the slightest breeze; it seems unlikely that celestial bodies of the same relative sizes and distances could ever negate the expansion of space, especially since they tell us that new space is added all the while to maintain density. (Yes, that's what they say.)

One of the assumptions for the Standard model, you remember, is that "the densities of matter and energy have remained almost uniform throughout the universe, and that any changes in matter or energy, as time went on, were extraordinarily smooth." Well, ...the question, "How can the density of matter in the universe remain uniform throughout the universe considering the fact that the volume of a sphere increases with the cube of the radius; and since each increase in distance results in a much bigger volume than the last increase in distance?" ...is dismissed with the handy notion of the constant introduction of new space. Perhaps a pudding the size of the universe might approximate a large-scale expansion, but, there is no source for adding more pudding, and there is no exterior heating! And, ...there is no pot! ... There is nothing against which we can measure speed; no fixed reference. Any measurement in relation to a neighbor would be the same, and all speeds would be *relative*, but those parts of the pudding furthest from the center would certainly not have the same density.

In our universe, if redshifting is reliable, it is measuring the distance of galaxies and their relative speed. Right? Maybe not. It is generally agreed that we can see just a tiny fraction of the universe, but the farthest galaxies that we *can* see *already* seem to have a relative speed near the speed of light; and more distant galaxies *can't* be going much faster. Since redshifting, then, can't be related to the motion of bodies much farther out, we have only two choices:

a. Redshifting is not a reliable measure at long distance, or...

b. The pudding is not expanding at a distance, which means that the universe is not uniform, and cannot be the result of a Big Bang; and redshifting, then, would be irrelevant.

Redshifting Discussed:

Two scientists have discovered an odd couple near Virgo consisting of the brightest quasar ever observed (at that time) and a newborn galaxy. They seem to be connected by an electromagnetic jet from the quasar that may have given birth to the new galaxy. What's odd about that? Just that the quasar has a redshift about 40 times greater than does the cloud![77] But, one scientist has proposed that the spectral pattern of a light beam should vary with distance as it travels through space, depending on the coherence of the millions or billions or individual microscopic radiators in the source.[67] Apparently the coherence of these many contributors to the light beam vary in their relation, and an observer will see a different spectrum *depending on distance*. This effect, verified by experiment, seems more applicable to non-thermal sources, and quasars are considered as such! So, not to worry now, ...it turns out that redshift/distance relationships may not apply to quasars.[170] Maybe so, but, this, of course, is another indicator that redshifting may not be an infallible indicator of distance, especially long distances, as we just discussed in the pudding recipe.

It is observed that:

> Ordinary galaxies have an intrinsic redshift that is ...correlated with younger age, and, "...the smaller and younger the object, the higher its intrinsic redshift."[249]

> "All of the smaller galaxies in our Local Group are redshifted with respect to the central galaxy."[247]

> "Smaller companion galaxies are systematically redshifted with respect to the larger galaxies."[248]

> "...That high redshift galaxies are at the same distance as low redshift galaxies – has been proved so many times before... and rejected each time."[250]

> "...The youngest stars in our nearest neighboring galaxies, the Magellentic Clouds, were almost all intrinsically redshifted."[251]

> "Even Stars in our own galaxy, the brightest, bluest, youngest, show that they are systematically redshifted."[254]

In addition, at least one scientist has noted over many years, that redshift seems to be related to the *type* of galaxy. Spiral galaxies redshift more than elliptical; dim galaxies redshift more than bright. More momentous was his discovery that redshifts differed from galaxy to galaxy in discrete amounts and that they might be *quantized!* This means that redshift could be an inherent characteristic of the galaxy, and that speed may not be the only determinant! Not only that, he discovered that individual reshifts may change with time! Needless to say, this blew the sneakers off some scientists, and a study was quickly undertaken to discredit such nonsense. Much to their surprise, the study verified the concept of quantized redshifting. One model of the universe, the Meta model, (We haven't mentioned this one yet, have we?) holds that redshift is caused by light passage through a denser medium and not by expansion.

> "One point at which our magicians attempt their sleight-of-hand is when they slide quickly from the Hubble, redshift-distance relation to redshift-velocity of expansion."[34]
> "There are now five or six whole classes of objects that violate this absolutely basic assumption. It really gives away the game to realize how observations of these crucial objects have been banned from the

telescope and how their discussions have met with desperate attempts at suppression."[34]

Antimatter and Matter Discussed:

Because gamma-rays don't seem to be very prevalent in space, as discussed a bit ago, the odds against finding much antimatter in the universe are not very good. But, we have been left some loopholes. The reason as explained to us for not seeing many antimatter particles in our galaxy lately, is that statistically they would have had to do all their colliding with matter long ago. And, if we take that reasoning to an antimatter galaxy, any matter particles in that galaxy would have had their collisions long ago, and again, there would be no gamma-rays to detect; even if – especially if – that galaxy was entirely made of antimatter; and, ...since galaxies are made all at once, any one, ...and perhaps a lot of them, ...could be antimatter.

The point of what I said a while ago, "Apparently there can't be much antimatter in the universe at all, unless it is separated from matter by the equivalent of galaxy cluster separation;" relates to the concept that if an antimatter galaxy is sufficiently distant from a regular matter galaxy, we would never know it, because they are then not inter-acting, and you can't tell by looking. The door left open by galaxy-cluster-separation distances between matter and antimatter galaxies is still open. It is indeed possible that some of the galaxies that we observe, or a lot, or even the 50% that we need, may well be antimatter.

And, there is another door open in the conundrum of the specified matter/antimatter ratio, that of the illogical reasoning of the Big Bang scenario itself. The very fact that matter and antimatter are self-annihilating reinforces the realization that they are contradictory. No one force, or propensity, influence, or proclivity; not even all of the king's horses, could ever, ever allow two such unequivocally opposite substances to be formed together, with each other, in the same place. You just can't boil and freeze water at the same time. Period. And if that first instant of force had a physical propensity for the creation of mass, how could the very next instant of that same force have the opposite?

A desperate retort might be that "the instantaneous nature of the event, the immediate conversion of the disparate forms into energy, begs the question." No. The two different materials could not possible have been made together in the same place by any one event, force, or imaginative consideration. The open door of galactic separation, then, is a more viable option than a Big Bang happening.

As far as the basic determination that symmetry demands that the two materials be made together, there is no way that I can comment on its correctness. (We can question its applicability, though.) I do read that there are aspects of symmetry that have yet to be tested, and if so, perhaps it may be later found that the two materials do not have to be made simultaneously after all. Or, ...or it might be found that the law allows the creation of the two materials, simultaneous perhaps, but in two very widely separated areas, such as in two separate galaxies. I am much more comfortable with this open door, even if it is a bit narrow, than I am with the exotica of Big-Bangdom.

And! There IS compelling evidence for that "open door of galactic separation" argument. Until recently, there WASN'T any evidence of mater/antimatter collisions in our galaxy, but now there is![241] It isn't discussed much, because it must be embarrassing, but a significant observation of some temporary gamma-ray indicators of antimatter presence was detected in the direction of our galaxy center in 1974. It didn't last, but just a while ago, (1997) "huge jets" of telltale radiation were discovered in that same area! And... wouldn't you know... they are blamed on a black hole.

But! There are a couple of ways that such an observation can be explained, and we don't need black holes. One rational justification for such a display is that some antimatter material – perhaps a few lone stars – is passing through. This IS embarrassing for cosmologists, because it is marvelous evidence that antimatter, separated from matter, IS roving the universe. And, if that is so, the argument, of matter – antimatter and annihilation, so crucial to the Big Bang theory, simply falls apart. And, ...there is the possibility that "annihilation" is an ongoing event in the development of galaxies! We'll talk of that later, also.

The Microwave Background Radiation Discussed:

As early as 1926, a temperature of outer space "produced by the radiation of starlight" was proposed, but after the Big Bang conception, "starlight radiation" was to be now called "Microwave Background Radiation." The now detected microwave background radiation is said to be proof of the original annihilation of matter and antimatter that the theory dictates: that when matter and antimatter destroy each other, their entire mass is converted to gamma-rays. In the case of the Big Bang, *two billion times the mass that remains in the entire universe today* was to have been converted to gamma-rays in that matter-antimatter merge, and that radiation is supposed to have cooled and redshifted over the billions of years back into the longer wavelength, lower energy microwaves that we observe today as the microwave background radiation.

Gamma-rays, remember, are those fearsome, harmful high energy nuclear rays that we all dread. Their wavelengths are shorter than one billionth of a centimeter, while microwaves are between one centimeter and 100 centimeters long. Microwaves, then, are anywhere from one billion times longer, to 100 billion times longer than gamma-rays, which are supposed to have redshifted up from the atomic radiation wavelengths, through x-rays, ultra-violet light, visible light, infrared light, and into the microwaves which we use in our kitchens. This has to be extraordinary, because the fastest moving galaxies ever detected, which are on large, assigned the same ages as the microwave background radiation, shifted only from normal visible light *towards* the infrared, just a few percent.

The now detected radiation seemed too smooth, so the COBE satellite was launched and its results celebrated. But the finally measured 1 in 30 millionth variations were thought to be too weak, and the patches of sky at the same temperature were thought to be far too huge, to have permitted the galactic and structural fallout. The MAX balloon observations were more accurate and smaller patches of equal temperature were found, so Big Bang is said to be confirmed.

But there are other considerations.

The COBE and MAX results didn't jive with the much smoother background required by the prevailing cold dark matter "theory;" hence the origin of the cold *and* hot dark matter theory, which we'll talk about shortly. Some scientists believe that galaxies came first and the microwaves second: that they are unrelated; that the radiation is too smooth and was more likely caused by cosmic dust problems.

"It is normal radiation from stars, which has been scattered, absorbed, and reradiated."[48]

One Scientist argues that infrared and microwave signals from ordinary emitters experience absorption, and the so-called background radiation could not have remained uniform, and would have been absorbed and unable to reach us over such long distances; that what we are observing might be more local.[124]

Some scientists feel that the background radiation is a scattering of ambient radiation by small filaments of plasma. A couple of scientists at Los Alamo National Laboratory have demonstrated that electrons trapped in strong plasma filaments could emit microwave radiation. It has also been demonstrated that filaments emanating from galaxies emit such radiation continuously. In other words, it is suggested that there is more than one way to generate these microwaves. Black bodies, for example, were around long before the Big Bang was invented, and "any source in thermal equilibrium with its environment, being neither heated nor cooled by gaining or losing radiation," will radiate a black body spectrum. Of course, there are all kinds of different radiations that permeate the cosmos. Cosmic rays, emanations from birthing galaxies, supernovae, pulsars, what have you; all adding up to the stuff of the universe and a perfect black body radiation of 2.73°K.

We said a moment ago that the detected patches of equal background temperatures were far too huge to have generated galaxies and such. What was meant is that across any area of equal temperature, there are no differences in potential. We ordinarily use this term in reference to voltage, but more broadly here, we can say that there is no evidence – it's an impossibility, actually – that any one such patch of equality could have produced *anything*.

It follows then, that it could be the boundaries themselves, and only those boundaries between the equal temperature patches, that could evince earlier galactic precipitation. But these boundaries are clearly thin and very long if the patches are so big; and galaxies formed along such lines would be grouped to look as if they formed a universe-encompassing chicken-wire fence instead of the uniformly peppered canopy that we experience.

The night sky is just chock-full of billions of billions of those galactic dots spread evenly everywhere. If an energy wave had gone through that area and formed all of those tightly packed pixels of light, no variation in the residue of its energy or temperature levels could possibly remain. Electromagnetic waves, after having gone through a suitable grid or mesh suffer refraction, remember? And this, in effect, is what we are discussing here. That original wave, if it had caused or gone through the fine grid that all these galaxies represent, would have seen all of its lobes of temperature differences change direction; and mingle and intermingle, over and again, so that any temperature differences would have averaged out almost instantly! Why do microwave relay stations use specially designed dish antennas? Because there is a natural spreading and weakening of the beam. *The very differences that COBE sought, found, and celebrated, are evidence of their theory's failure!*

Even if we disregard the refraction argument, if the temperature differences are not natural and infinity old, and are indeed the result of an event; an averaging process would have to be on-going. Any differences would disappear after a few billion more years leading future scientist to infer the opposite of what our here-today scientists have concluded. Is it just a lucky coincidence that we just happen to be here just as the background radiation showed the "proper" temperature deviation?

And more...

You know, mass represents a *lot* of energy. If two beams of energy are traveling together, side by side, and one of those beams gives birth to a galaxy on route; that beam is going to have a *lot* less energy (temperature) than its fallow companion does. And even if these beams did not mingle in the throws of refraction, then there should be temperature differences more significant than 30 millionths of a degree.

And, energy rushing off into space is undetectable unless it is reflected back from something else. If you shone a flashlight up to the night sky, what would you see? Nothing! There is nothing to reflect the light back. Since the scenario requires that nothing be out there before the gamma-rays arrived, nothing would be reflected back. If we are told that the radiation actually emanates from clouds of the so-called primordial residual material that didn't form into galaxies – rather ordinary dust – then gamma-rays and the matter-antimatter merge are non-issues.

Background Temperature Discussed:

Finally, we said a moment ago that the temperature of that perfect black body, "stuff of the universe radiation," is measured at 2.73°K in keeping with what the Big Bang theory predicts. (Some of cosmology's favorite phraseology has to do with the Big Bang "predictions," ignoring Earthly constraints on the language. The rest of us can only predict future events, but cosmology feels no such compulsion.)

Cosmologists estimate that the Big Bang happened between 15 and 20 billion years ago, and from the official, initial temperature of 10^{27} Kelvins, as the scenario goes, the radiation of the expanding universe continued to cool. .. To 2.73°K. 2.73! There is something magical about that figure. To predict today's temperature, we would have to graph the change from the higher to the lower, over the period of elapsed time, from then to now; but if we don't know if the elapsed time has been 20, 15, or perhaps 8 billion years, how do we graph it?

Supporters were happy with the 2.73 figure back when the universe age estimates were between 15 and 20 billion, so why doesn't the more recent, lesser estimates of 8 billion years result in a higher temperature for today?

Of course I'm being picky. But everything about the Big Bang invites such nit-picking, because everything said in cosmology is one bit of nonsense after another. 2.73°K is darned near absolute zero. Right? For the fun of it, run a test of your own:

> a. Explain to a friend that 0°K is absolute zero, which is the temperature that molecules stop vibrating.
> b. Ask him or her, "…What would you estimate the background temperature of the deepest, most remote part of empty space would be if the Big Bang happened 100 or even 1000 billion years ago?"

The odds are great, if your friend stops to think about it, that the answer will be something like: "Oh, …a couple of degrees above zero, I guess." And if your friend can come up with that answer, what does that say for their academic, scientific "prediction?" There is *no* way that the temperature could be very much different than it is, …regardless of the universe's age, … because we can't cool the universe to absolute zero, no matter how long we take. It really is, …after all, "the radiation of starlight."

Light Element Ratios Discussed:

Estimates of the universal abundance of light elements are certainly based on a deep understanding of particle physics and sophisticated mathematics, and they are well supported by good science; we respect such analysis. But, hydrogen, which is thought to be made only in the extremely high temperatures of a Big Bang, is also the most abundant element in the universe. That bothers me. You and I, if we are not scientists, probably would have guessed that the lightest, simplest element would be the easiest, the first, the most prevailing for nature to make, and because it is the most stable and most durable, the most abundant.

It's not usually very productive to question good science, but whenever a magical or mysterious reason is offered to explain the ordinary, we can't help being a little suspicious. Our experience and maturity dictate hesitancy when we are offered unworldly answers. The super-abundance of hydrogen, for example, calls for a routine, plain, …mundane process of manufacture. I would rather have it born in stars, or have its quarks and leptons precipitated out of energy fields like virtual particles (new material); I would rather have it that science may not completely understand what happens deep in the bowels of stars than to require a one-time unbelievable happening. Before the Big Bang was invented, scientists did not seem perplexed with observed light element ratios. Why are those ratios so unsettling today?

It is well accepted that nucleosynthesis in stars still goes on today; that the abundance of elements and their ratios in these stars depends on, and varies with, the size, age, and other characteristics of those stars; and since we can never make actual measurements or truly test the theories involved, it seems myopic to discount the reasonable guess that what we are observing might well be the product of day-to-day ongoing stellar activity in average, ordinary, normal stars. There is no evidence whatsoever that the observed abundances aren't the result of an infinitely sustained, natural production.

Including the manufacture of hydrogen! Quasi Steady State model supporters, at least, don't believe that Big Bangs are necessary for such manufacture. Big Bang believers say that any old supernova can form all of the other elements in the universe, but that we need super-high temperatures to form hydrogen, in spite of the fact that hydrogen is the simplest and easiest to make. Just let an electron get near a proton and presto: a hydrogen atom. And, ...it was recently reported that a couple of scientists have discovered a new class of very dim – but very big – type of galaxy that *outnumbers* those with which we are most familiar. And ...that they are essentially "enormous disks of hydrogen gas that are massive enough to outweigh normal galaxies!" What does that mean? Haven't we just *doubled* the amount of earlier observed hydrogen? What about the estimates? Also, it should be noted that the temperature at which these gases were to have formed happens to match the temperature of star interiors!

Deuterium, "heavy hydrogen," is also said to be made only in a Big Bang scenario; but to calculate what that abundance might have been, it turns out, scientists have to guess at the rate of star birth, and a guess is all that they have. The ratio of deuterium to hydrogen has been a key determinant for the Big Bang theory, but physicists can only make rough estimates of the ratio because it would depend on the total density of matter at the time of the Big Bang, and it turns out that deuterium is destroyed in stars in the process of making helium. From the amount of deuterium implied by those recent measurements of the old gas clouds, extrapolations are made back to the Big Bang to determine amounts, but their measurements may have been inaccurate because of ...interference. Recent measures of its ratio to hydrogen indicate the density of mass might be five times higher. A "less likely alternative" is that "the Big Bang model might be incomplete or wrong."

Recent Mount Keck measurements, by the way, indicate that a "surprising abundance" of deuterium has been detected at great distance) (say age), but (handily) "the ratio is still within the limits predicted by the Big Bang." Well, ...what did they surmise? Not that star birth estimates might be wrong, but that there needs to be three times more dark matter than Big Bang theories have needed up 'till now![143] The frequent use of exotic phenomena to excuse observations that don't back up the theory seems like a knee-jerk reaction, but it happens continuously! Wouldn't it be more prudent to question estimates of star birth instead?

Helium, the second lightest, simplest, should indeed be the next most abundant, shouldn't it? It is made in old massive stars,[69] and the older they are, the more helium they can make. Indeed, a lot of medium stars, if not most – if not all – are "known" to have solid helium cores. (Though, as a star evolves, its core tends toward the heavier elements, with iron predominant in the most massive.)[286] Couldn't the observed ratio of hydrogen and helium be anticipated *regardless* of how the universe was formed? Some scientists believe that there is more helium than could plausibly be made in existing stars, but what about all of those very large, very dim galaxies mentioned a moment ago? What about all the stars that have come and gone throughout the eons? And, an estimate that observed stars by themselves are not enough to have formed all of the observed helium is ignoring the realization that we might not understand stellar creation perfectly. (Where are all those neutrinos that are forecast?) And, there are many, many more stars than have been observed up to now; more and more are being discovered daily; exponentially more.

> "It rests, however, on many untested, and in some cases intestable, assumptions. The result is a bandwagon of thought that reflects as much faith as objective truth. The Big Bang belongs in the realm of metaphysics, not science."[6]

Lithium and deuterium are widely understood to be burned in low-mass stars, which might explain the apparent low abundance of these elements, but those elements might well reside in brown dwarfs. Brown dwarfs, which are thought to be too small to attain nuclear burning, are sought in the hopes of verifying the presence of lithium, but since they don't burn like stars, they are very dim and very difficult to find. One or two candidates seem to have been spotted, but their estimated age raises more questions than answers about the material's supposed depletion process. Brown dwarfs, like most other non-burning material in space, are almost impossible to find, so estimates of their abundances are just that. Lithium, then, and perhaps boron or deuterium, might never be accurately inventoried. Beryllium abundance is about 1000 times that predicted, ...etc.

Element abundance "predictions" don't seem to work out very well, so "adjustments" such as increasing deuterium destruction, offering an inhomogeneous Big Bang in one case, while suggesting a very homogeneous Big Bang in another, are tried. Cold fusion may have lost its charm, but for a while, many scientists embraced its possibility and thought that they had even observed it. Their enthusiastic over-reach is perfectly understandable, but it is a tacit admission that these processes have unknowns yet.

"...these observations pose serious difficulties for those wedded to a uniform explosive beginning."[17]

"Certainly, physicists are a lot further from understanding how the world works than some would have Congress believe."[21]

"Often, the scientist or visualizer has no intent to deceive but merely wishes to make an attractive picture or to present a compelling argument."[113]

Mark Twain's *Life On The Mississippi,* is quoted: "Science is an endeavor in which one gets such wholesome returns of conjecture out of such a trifling investment of fact."

3 June, 1991   *Newsweek* p. .51,

Chapter 9

Dark Matter

The most famous Big Bang inventions, of course, are the hot and cold forms of dark matter. In yet another of those cases where exotic phenomena are used to excuse observations that don't back up the theory, we are told that a unique (say peculiar) decay process (that would require fanciful types of fermions [mass] and bosons [force carriers] that didn't react with the rest of the bunch, [add these to the list in Chapter 13] could have allowed the two quite different types of dark matter to form together.

Yes, if we are going to have "cold" dark matter, if follows that we should also have "hot." Cold has been the more popular concept of dark matter, but, ...the standard cold dark matter model couldn't explain all of the observations, so... hot dark matter (massive neutrinos, they say) was invented. Cold dark matter is supposed to move slowly and form structures only gradually, but hot dark matter is said to move at nearly the speed of light, form super-structures first, and only then break up into smaller galactic organizations. Wouldn't you know it though, each type would have to develop entirely differently,[89] and we would have to invent new types of particles that haven't even been considered up till now. And... since observations don't seem to fit either model well, ...a team of physicists suggests a mix of 2/3 cold dark matter and 1/3 hot dark matter![111]

Our now old argument that there could be only one compulsion to make things in that big Bang has to apply to different types of dark matter and imaginary particles as well. If matter and dark matter were spread throughout the ball in an homogeneous soup, a fixed ratio of matter and dark matter would have to exist everywhere, and any dark matter gravitational force would probably be self-canceling as far as galaxy formation goes. If they were formed together, and dark matter's only contribution to the universe is a "needed" gravity, wouldn't nature have simply given mass more gravity?

One of the predictions (say requirements) of the Inflationary model of the universe is that all the visible mass amounts only to about 1% of the total matter, and the other 99% is dark, invisible, ...different; emitting no radiation at all, it is inferred only through the motions of galaxies and the "need" for something to help "close" the expanding universe, and it is called dark matter. It's "dark" because it can't be detected. Of course, there is a lot of ordinary matter out there that is difficult to see, including planets, dead stars, neutrinos, dust, and similar "normal" debris; this was the reason, you remember, that Quasi-Steady State universe supporters don't require dark matter. They say there is plenty of this dead stuff lying about. Certainly there are stars and galaxies that we can't see – yet – because they are simply too dim for present equipment. Supposedly black holes are out there too. But, dark matter includes none of these; it's ...*different*.

There most certainly are plenty of stars and galaxies that we haven't seen yet: they are being discovered all the time. Each time a new, more sensitive instrument is employed, a plethora of new galaxies come into view. A 1990 article in *Nature* noted that the vast numbers of galaxies being found already poses a big challenge to conventional ideas.[53] Then, in July of 1991, *Discover* magazine reported that one scientist has discovered billions upon billions of previously undetected galaxies filling the sky, such that the full moon hides over 50,000![35] And, as we said earlier when we were talking about light element ratios, that same magazine reported in May of 1994, that two scientists have discovered that dim galaxies, five to twenty times dimmer than normal galaxies, but many of them much bigger – one, the biggest spiral galaxy known – may *outnumber* the brighter galaxies that we are more familiar with. Additionally, astronomers are *still* missing a lot of galaxies: dim "dwarf" galaxies that may outnumber *all* other types of galaxies. The telescope at Apache Point in New Mexico is expected to detect *100 times* more galaxies than we have so far discovered. And, ...new, more powerful telescopes are coming on line, one after the other. No one has remotely suggested that we are seeing all that we can see; that we are approaching any limits, or boundaries, ...or edges. There is every reason to assume, and anticipate, that each and every more sensitive search will sight more...and more...and more, yet the stupendous number of galaxies already found poses big problems for Big Bang believers.

What Is Dark Matter?

But, as we said, dark matter includes none of these "normal" objects, it is ...different. Dark matter is believed to be different from anything like the matter that we are familiar with. Supposedly it is not the dying galaxies, or rocks, or dust, or ancient debris that we envision. Call it transparent, invisible, or strange because it doesn't react with ordinary matter, or with magnetism or electromagnetic energy. It can pass through you and I, the Earth, and we wouldn't notice. ...Wait a minute. Dark matter's only effect is gravity, which would tie it to observable mass, which means that it can't move through us or in relation to us at all! Nothing in the visible universe fits the description of dark matter, but it would have to have mass and it would have to be stable. It's bizarre, weird stuff, little black holes, or something we haven't even imagined. Funny, though, cold dark matter theory holds that variations are the same everywhere, but string theory holds to the very opposite.

Dark matter is said to be non-baryonic. Baryons are constituents of regular matter: ...you know ...protons and neutrons and stuff like that there. Dark matter is not made or protons and neutrons – it's not "real." It's invisible, it doesn't seem to be effected by anything, and it seems to effect regular mass only weakly, and only by gravity. *Two types* are envisioned: one that congregates around individual members of a cluster, and a type that gathers around the cluster as a whole. There is a number of suggested "forms" of dark matter: axions, cosmions, cryptons, champs (charged massive particles), higginos, machos (invisible small stars), neutralinos, photinos, WIMPS (weakly interacting massive particles), winos, and zinos. And, ...a massive version of the otherwise very tiny neutrino is hopefully offered

Arguments for Dark Matter:

The Big Bang theory *demands* the existence of dark matter. Because the universe was supposed to have been extraordinarily smooth at the "creation," it is determined that the observed very large structures of galaxies that are spread over 500 million light-years, and even the presence of galaxies themselves, could not *be* without the help of additional gravity. A *lot* of additional gravity it turns out; about 100 times more than can be had with the mass that is visible in the universe. We have been told right along that perhaps as much as 99% of this handy material is hiding in massive halos surrounding our galaxy and others, but now, those recent Mount Keck Telescope measurements indicate that there might be (needs to be) three times more dark matter than earlier estimates, and ordinary matter amounts to only about 1/3 of 1% of the total mass of the universe!

The basic problem here is that the argued 8 to 20 billion years since the Big Bang, is nowhere near enough time for all the clumping and lumping that we've had, because the visible matter couldn't have generated the gravity needed. So ...it follows that there must be one heck of a lot of invisible matter out there, or the Big Bang simply wouldn't work. One proposal has it that dark matter may have formed *earlier*[47] than ordinary matter did, helping to form points of higher density that would have permitted matter to eventually develop the large-scale structure now observed.

Also, most galaxies are seen to rotate too fast; and some have huge, otherwise unexplained, distortions. Scientists haven't been able to explain how galaxies could rotate as fast as they seem to do without breaking up, because the amount of visible matter doesn't seem anywhere enough to hold things together. Without dark matter to make up the difference, gravity couldn't hold the stars and galaxies together. Even pairs and groups of galaxies seem to move around each other faster than visible matter allows.

You probably remember that cold dark matter was supposed to help "close" the universe, or at least help keep it "flat." Scientists feel strongly that the universe is not open, that it undoubtedly will either stop expanding and close back up into a raison again, or it will gradually slow down and come pretty much to a halt. Both of these scenarios demand a *lot* of dark matter.

In support of gravitational considerations, particle physicists have pretty much determined that there are only three families[58] of fundamental sub-atomic particles. You are undoubtedly familiar with electrons and neutrinos, but there are a host of others with strange sounding names: up and down quarks, muons, taus, and ...strange; all, however, grouped into only three families. Considering the three-family limit and the observed abundance of light elements, scientists estimated how much baryonic or "normal" material there is in the universe, and decided that we need a *lot* of dark matter.

Evidence for Dark Matter:

The smooth but slightly variable microwave background radiation detected by the cosmic background explorer satellite, COBE, is said to support the dark matter theory, because there is not supposed to be any way that the background radiation could be as smooth as it is without an ameliorating force such as would be provided by dark matter.

The ROSAT x-ray satellite discovered a cloud of gas with more mass than 500 billion Suns that has a temperature of almost 20 million degrees. This is taken as "evidence" that an immense concentration of dark matter, with its powerful gravity, *must* be holding the cloud together. And ...one galaxy is thought to be distorted by some kind of unseen powerful force; dark matter to be sure.

One scientist has "seen" it. Arcs of light, in the "lensing" effect forecast by Einstein, bending around apparently invisible mass about the size of a galaxy in the center of the galactic cluster. The way he did this was to "factor out" the distortion "likely" to be caused by nearby visible stars and galaxies, so that he could calculate the density and distribution of dark matter responsible for the remainder. The invisible dark matter seemed to be concentrated in the center of the cluster, gradually becoming much thinner as it spread to the edge. What he has found is considered indirect evidence of dark matter. Other observations seem to support the idea of such an undetectable glob in the center of other clusters of galaxies.

An American-Australian team, after studying over three million stars, reported that they had detected *one* convincing occurrence that could imply the presence of machos. They reported that a star in the Large Magellanic Cloud began to brighten suddenly in January of 1993, became seven times brighter by March, and then gradually return to its original brightness. They concluded that they had witnessed a microlensing event. The object that would have caused it was thought to be on the order of 30% the mass of our Sun or smaller. A French team made a similar observation at the same time. There is still the chance that what was seen could have been a variable star, but since the French saw the same effect in a different body argues against that possibility.

Arguments Against Dark Matter:

Dark matter is not a prediction. Predictions are made before, not after. Dark matter is an invention; deemed absolutely necessary to make the Big Bang theory work: it is in every sense, a desperate last-ditch effort to keep the Big Bang scenario alive, a series of one exotic invention after another. Just as scientists were at one time convinced of the existence of the aether, the "necessary" medium for light to travel in – to the point that they even determined its density – they are today convinced of the existence of dark matter and have analyzed it to the extreme. Einstein never rejected the aether, but since he determined that it could never be detected, he felt that science needn't waste its time looking for it. With his guiding logic, since we can't detect dark matter, perhaps it doesn't make any difference whether it exists or not, and we, perhaps, should select a different model. Since Einstein has enough rules named after him, for the fun of it, let's name this: "Roy's Rule:"

"If it doesn't matter, forget it."

Cold dark matter is supposed to have started clumping first to help with the building of regular matter. Without it, galaxies couldn't hold together; they spin too fast for their observable mass, remember? But if you were tasked to solve this little problem, wouldn't you first consider more worldly possible solutions before evoking the exotic? For instance: Is magnetism a possibility? What about electric charge? Is it possible that our measurements of speed are in error, or our calculations of the amount of necessary mass? How much more normal but "too dim" material is out there? And, how about antimatter? Do we *really* understand galaxy development? Scientists have, of course, addressed these possibilities, but they seem too committed to one, unlikely, scenario, and they do seem far too eager to accept the "unreal" to explain the real, especially since there are other concepts to consider.

It seems that just about all of the dark matter in the universe has the same distribution as regular matter, to the extent that where you have visible matter, you are supposed to have even more, much more, dark matter mixed in with it. Scientists, for example, calculate that there is much more dark matter in our own galaxy than there is regular matter, and it resides in a halo "at least 6 times" bigger[82] than the galaxy that we can see. Now, the way they determined this was by analyzing the motion of the Large Magellanic Cloud near our own Milky Way. The trouble is, it is very difficult to make such measurements and their accuracy is questionable. For example, the brightness of Abel clusters (a type used as standard yardsticks) was determined to be different depending on where we looked, and the difference was attributed to our own galaxy's motion. Equally, a difference in the appearance of the microwave background radiation was also credited to our own motion. However, the differences do not match what we should be seeing for the velocity we are said to be experiencing. Not only that, at least one scientist said that there is no absolute reference against which we can make such measurements, including the microwave background radiation, and, "perhaps ...nothing is moving at all." [68]

Well, since the measurements of motion that seem to indicate the presence of dark matter might be quite erroneous, perhaps such measurements should not be relied on for inferring the presence of exotic material. If we could see disturbances in galaxies that could only be blamed on some unseen mass, perhaps the concept of dark matter might be a bit more viable, but, for the most part, the galaxies seem quite undisturbed by anything at all. Of course, we did cite, as evidence for dark matter, the "lensing," or bending of light rays that seemed to indicate a dark matter mass, and the temporary brightening of a star that was also attributed to something invisible; but neither of these events are hard evidence, especially since more reasonable reasons could be posed.

For example, to obtain the lensing effect, "likely" distortions due to other nearby bodies had to be guessed at and factored out. Nothing wrong with the procedure, but cosmology is by nature a series of such educated, sometimes ingenious, guesses that may or may not be correct; indeed, how would we know? As for the temporary brightening observation, there was no mention of any dimming that would necessarily have to proceed, and then follow, a brightening that could be credited to a passing body. If you picture a gravitational body approaching a light beam coming from a distant star to the Earth, you can see that the intruder would have to first bend the beam away from its original path (dimming), before it could get right in the beam (amplifying); and then as it was leaving, bend the beam away again (dimming). There was no such observation noted. On top of that, the amplification can happen only when the object is exactly in the beam, but the dimming would happen if the object only came quite near the beam – a much larger statistical likelihood – yet we have no such events reported.

Typically, dark matter is pictured as infiltrating and surrounding a galaxy even out to its far environs, It is supposed to be invariably associated with mass, indeed it is by nature attracted to mass, yet; in spite of the fact that it is supposed to be as one with mass (galaxies), and in spite of the fact that galaxies should result in blurry lensing, the images shown as evidence of dark matter are sharp. And ...in spite of the fact that you are not supposed to have one without the other, no image was seen where the dark matter was supposed to be. So ...the *absence* of observed matter was held out to be proof that the lensing was due to dark matter. This is a beautiful example of "selective justification." On the one hand dark matter is supposed to be coincident with regular matter, but when regular matter can't be seen, its absence is taken to be proof of dark matter!

Seldom are the more mundane explanations considered. For example: Could the effect have been caused by a neutron star? Or a black hole? There have been precious few examples found of lensing arcs, but numerous examples of "optical illusions" showing two or more seemingly identical bodies when in fact they are different images of the same one. Some kind of illusion might be involved. We have seen a number of images with rings in them, supposedly formed from the central star's explosion, even though a ring is much more odd than a spherical shape. Could whatever formed these rings be forming the so-called lensing examples?

Perhaps we should also consider the statistics involved. We are told that there is 100 to 300 times more dark matter than matter, yet after countless studies involving millions of bodies, only a handful of lensing observations have been attributed to possible dark matter objects.

"The study is ...probably telling us that we don't fully understand [gravitational] lensing."[90]

Doesn't it seem to be a coincidence that matter and dark matter are usually in, even have to be in, the same locations? Couldn't this be an indicator that basic measurements might be wrong? We are told that there is not enough visible gravitational matter to allow all this clumping and structuring in a mere 15 to 20 billion years, so dark matter was invented. Couldn't this be an indicator that our age estimates might be wrong, ...or even (blasphemy!) our understanding of gravity? Perhaps it isn't the motion of galaxies or time restrictions that require the presence of dark matter so much as does the myopic acceptance of the Big Bang model itself. Remember, even Einstein allowed himself to make a mistake because of the prevailing prejudice of the aether; now we have another debilitating dogma.

How can dark matter *not* radiate? As regular mass compresses under gravity, it is said (theorized) to emit gravitational energy as radiation. Supposedly galaxy-forming clouds kept condensing and fragmenting in stages until each fragment became too dense for the radiation to escape. Well, that escaping radiation was to have resulted from the process of gravitational compression, whether the mass was visible or not! According to cosmologists themselves, it is inescapable and inevitable that there be radiation in such a scenario. How could any matter in this situation, dark or light, not radiate? On the other hand, if gravity doesn't radiate, and there is no evidence that it does, then the prevailing concept of galaxy development needs revisiting.

Theorists need dark matter to form galactic structures, but it's possible that short cuts and simplistic solutions to Einstein's equations might be causing the difficulty. The equations are non-linear and very complicated, and they are still being studied today. The newer studies of chaos (a recent, popular mathematics discipline) show that only a few variables can render postulates impossibly unpredictable, and Einstein's theories have plenty. Further study and more comprehensive solutions might negate the need for exotic invention.

We are told that the cosmic background radiation had an homogenizing effect on ordinary matter, but not on dark matter, and that dark matter could have begun clumping "long before" regular matter could have. In spite of dark matter's only reputed similarity to regular matter: gravity, dark matter was allowed to clump and ordinary matter wasn't; yet "clumping" is a gravitational activity. Why they would be treated differently is unexplained, as is the ultimate breakup of this primordial (unreal) clump. And, if *anything* happened "long before" matter started to be, doesn't that cut down on the time that matter had to form galaxies? They were already time-crunched before this problem came to light.

Evidence Against Dark Matter:

Scientists seem to feel that WIMPS are the most likely dark matter particle dominating the mass of the universe, but none have been detected; and more and more qualifications, restrictions and stipulations are offered. Axions and neutralinos have been arduously (and expensively) searched for, and still are, but to no avail. In a strong magnetic field, they are supposed to convert to photons, but calculations suggest that there is no magnet in the world big enough to spot an axion, and there probably never will be. Still, they keep looking. And Machos: they just aren't a fashionable dark matter solution, and there aren't enough of them to "close" the universe, even if they do exist; and "closure" is required by some theories.

The Microwave background radiation was supposed to have proven the existence of dark matter, but as we discussed earlier, the observed, even predicted microwave background radiation can not be attributed to the Big Bang at all, and therefore is unrelated to any discussion of dark matter.

Astrophysicists have found that great clusters of galaxies surrounding voids hundred of millions of light-years across, are rather common in what they call "the great wall" and other structures. The immensity and expansiveness of these structures just doesn't square with the calculated distribution of cold (or hot) dark matter, and the theory is threatened. Twenty billion years aren't enough time to have developed these formations, even with the help of this invisible, exotic, hypothetical, even desperate embellishment.

Some British scientists have mapped a good part of the sky and report galactic lumpiness up to 150 million light-years across, and an astronomer at the University of Hawaii has found that our own galaxy is part of a super complex perhaps a *billion* years long. The extraordinarily lumpy universe suggests that galaxy formation has nothing to do whatever with the supposed existence of dark matter. Dark matter theories can't even account for structures over 30 million light-years across. These scientists suggest that researchers may have to abandon the cold-dark matter theory, which just can't account for such large-scale clumpiness.

The ROSAT x-ray satellite results, cited a moment ago, determined that a lot of dark matter, equal to approximately 20 times the observed normal matter, is required to hold that gas (plasma) cloud together. This "need" for dark matter was evidenced as "proof" that dark matter existed. Of course, this is no proof whatsoever, but 20% would be skimpy proof indeed, since we are told that dark matter equaling up to 300 times the amount of observed mass is required by the Big Bang theory.

One of the galaxies is thought to be distorted by some kind of unseen powerful force... But any such distortion would have had to have happened while the galaxy was forming, because dark matter is said to have formed before matter; it is the binding force that hold galaxies together and cannot effect one part of a galaxy more than another. Its effect certainly can't be transitory. And how can we tell that the galaxy is distorted and not in its original shape? They come in all shapes and sizes. And, why is this cloud not merely a galaxy in formation? Isn't this how they are supposed to be formed? Can we be certain that there's no regular mass in the cloud? We can't see through it. Why is the dark matter not tugging on the surrounding galaxies? How do we know the cloud is nearby?

Other scientists also realize that the concept of dark matter is losing its credibility. A few quotes:

"The theory is missing something profound."[16]

"Given that there is no empirical evidence for non-baryonic dark matter, and some evidence against it, should it still be our best candidate?"[52]

"The Inflationary model and cold dark matter were invented to save the Big Bang theory, but their help is in some doubt."[52]

"Both Inflation and Cold Dark Matter theories are vulnerable, and cosmology is in turmoil."[46]

"All the conventional dark matter theories are in deep trouble."[69]

"It's disturbing to see that there is a need for a new theory every time there's a new observation."[69]

"For the cold dark matter hypothesis to survive this crisis would take such complicated physics that the cosmos would have to operate like a Rube Goldberg machine."[47]

"There are alternative theories floating around – all within the concept of the Big Bang. They'll receive more attention now that people are at last convinced that one promising-looking idea [cold dark matter] is in deep trouble."[48]

"Strange lumps of matter on the out-skirts of our Galaxy may be stars from a bizarre 'mirror sector' of the Universe, say physicists in the US and Australia. Although these stars could be burning fiercely in the prime of their lives, the laws of physics that govern them would make them invisible to human eyes."
"Mirror matter would come with its own unique set of laws. It would feel the force of ordinary gravity, and could clump into mirror stars and planets. But its versions of the strong, weak and electromagnetic forces would be different from those we know. Although mirror stars would burn through nuclear fusion just like normal stars, they would not emit photons, so they would be invisible."[296]

Right now, the Big Bang and its dark matter invention can offer no explanation for the universe being the way that it is, whereas one or two of the less publicized versions of the universe do a pretty good job. We'll discuss those later, ...of course.

"Real"?

Well now, we have been told over and again, that 99% (or more) of all the matter in the universe is dark, invisible, that it emits no radiation; it doesn't react with magnetism or electromagnetic energy. Some of it moves slowly, the rest moves fast. It is admittedly bizarre, weird stuff. Unreal. In other words, you and I, and the Earth, the Sun, and every celestial object in the sky – the real stuff – are unlike the vast majority of the universe! Not only are we not at the center of the universe, we are not made of the same stuff? And we call dark matter "unreal?" Shouldn't 99% of the universe be considered real rather than the 1%? If majority rules, you and I can't be ...real?

* * *

Detective Columbo got out of the squad car and walked purposely through the door held open by an officer; and was guided into the study by Sergeant Creative. On the floor lay the victim with three or four stab wounds in his back.

"Have at it, Sergeant," said the Lieutenant as he took a cigar out of his pocket.

"Well, Lieutenant, the house keeper tells us that after realizing that the old man hadn't gone to bed last night, she knocked and called, and tried repeatedly to get in the room; but it was obviously locked from the inside."

"Is dat right?" said the Lieutenant, lighting his cigar.

"Yes, Sir, but get this. All the windows and the other door were locked ...from the inside. I don't see how our killer could even get out of here unless he could go through walls like a ghost."

"Well now, ain't dat interesting," said the Lieutenant. He scratched his head and puffed on his cigar, and after studying the scene at length, and just about when the cigar was nearing its finish, the Lieutenant walked out and gave tomorrow's headline to the press: "A GHOST DID IT."

* * *

The Cosmological Constant:

When Einstein developed the theory of relativity, he felt that a gravitational universe needed a "fudge-factor" to counter the effects of gravity; otherwise, everything would simply collapse on itself. When Hubble later declared the universe to be expanding, Einstein admitted that he had made a mistake and threw the darn thing out. ...But it wouldn't go away! With every latest measure of the rate of the supposed expansion, a more negative or more positive, bigger or smaller value was assigned to that bedeviling constant, ...and it has been confounding and tormenting cosmologists ever since. That factor was called the cosmological constant. It has been called: a repulsive force independent of matter to counteract gravity, a period of time, and popularly, the rate at which the universe is expanding (but no one can agree on its value – it is supposed to be expanding slower now than in the past.)

The cosmological constant is not a part of dark matter and is not usually discussed with dark matter, but they are related in that they are both intangible – pretty much indeterminable, but critical – invented underpinnings for the Inflationary Big Bang model. The cosmological constant is just one of the factors in the gravitational field equations – it is equivalent to assuming that empty space has a constant energy density associated with it – but it is used prodigiously, along with cold dark matter to dominate the expansion of the universe. It is conjectured that primordial dark matter would dominate the expansion of the universe for about half of its history, ...then the conceived force represented by the cosmological constant would take over! Theorists say that if 80% of the energy density of the universe can be attributed to the cosmological constant, it *might* explain the COBE findings regarding the microwave background radiation. It is crucially important to the story line, but it is the mathematical term that is giving scientists the fits.

As the theories of quantum physics and cold dark matter began to develop, it was realized that a cosmological constant of zero wouldn't work, because that would mean that the energy density of space is zero, and we know that space is chock-full of all kinds of radiant energy. Quantum theory not only holds that empty space is not empty, but that it is literally bursting with energy; so much so, that it amounts to 10^{80} ergs per cubic centimeter![31] (Remember, a billion is 9 figures to the left of the decimal point, and empty space had 80! This is a lot of energy!) Since Einstein showed us that energy is equivalent to mass, all this energy should be exerting a very noticeable gravitational force, perhaps to the point that mass gravity is essentially negated! Measurements do seem to indicate that the cosmological constant is either zero or darn close to it, so it is thought that *quantum theory just might be wrong in this case.*

But, again, the cold dark matter model requires a positive cosmological constant, which also bestows an energy density on the vacuum of space in an effort to explain how matter in an expanding universe could lead to the formation of the great structures observed. The model can explain structures up to the size of galaxies, though, but not much bigger. The trouble is, that if the cosmological constant was large and positive, everything would fly away; and if the cosmological constant was large and negative, the universe would ultimately contract back into that "raison."[31] So, it can't be *large*, but if it is zero, quantum theories would have to be wrong. If the cosmological constant could be determined as something other than zero, a lot of problems associated with redshift and the observed numbers of galaxies would be alleviated, but, scientists can't even agree whether the total energy in a vacuum should be positive or negative. Redshifts show no evidence for either, and many cosmologists feel that orchestrating the constant to save face is a last resort, because, a non-zero cosmological constant would have profound and disturbing implications for fundamental physics. Fortunately, one scientist claims to have resolved the dilemma with a theory that *wormholes* allow the universe to have a zero cosmological constant.[31]

This invented, later to be discarded, later to be retrieved fudge-factor has indeed been giving scientists the fits, because they need it, badly, even though it is nothing but a contrived bit of mathematical imagination! A "conceived force?" "If 80% of the energy density of the universe can be attributed to the cosmological constant,..." Now, what the Hell does that *mean*? Almost every cosmologist gives it credence, but what the heck is it? It's not gravity, it's not electromagnetic; it is a "conceived force" not considered by any scientists other than cosmologists!

And, Omega:

It is almost imperative for cosmologists, that the universe be delicately balanced between the squandered, anticlimactic, non-ending fade-out of an open, forever expanding model; or the "hot crunch" of a closed, collapsing scenario. Cosmologists characterize that balance with the term *omega*, defined as the ratio of the actual mass density to the critical density needed to create a flat, delicately balanced between open and closed, universe. With Einstein's value of zero for the cosmological constant, omega equals one, corresponding to a flat universe. A major reason that most astronomers favor the Inflation model, is because in this model omega equals one and remains constant with time.

Seemingly, most scientists are quite adamant that omega "almost certainly" has a value of one, in spite of the realization that an omega of one demands a lot of dark matter, because the very popular inflationary version of the universe and the cold dark matter concept crucially depend on an omega of one. But we are told that some empirical tests, and objective evidence, indicate an omega of no greater than 0.3, and that we might do better to put our trust in observation more then theory.[52]

It seems to me that a "delicate balance between open and closed" is equivalent to a universe that is not doing anything. And we don't need dark matter to demonstrate that. Since we can't arbitrarily change the gravitational relationship of mass, we might consider one other, more palatable alternative, one that is so simple and workable that it boggles the mind for its neglect – age. The argument that there was no Big Bang, and that the universe has been around for much longer, such as is claimed by Quasi Steady State advocates, eliminates the time constraints for forming galaxies and the need for an invisible, unreal dark matter.

Neutrinos:

The latest is that neutrinos do have some mass. Just a tiny fraction of an electron's, but enough, we are told, to provide all of that missing mass and gravity that dark matter has been providing.[279] The argument, though, is specious. There are far fewer neutrinos detected than theory calls for, and there has to be far more neutrinos outside of a galaxy than inside, so this new-found mass is of no help in keeping galaxies together.

Chapter 10

Galaxies, Etc.

Heavy Elements:

Cosmic rays don't get much press, but the Earth is bathed in them almost constantly. Though they represent just about every element found in nature, most of them are said to originate as the heavier nuclear weights; it is estimated that the lighter cosmic rays were originally heavy elements that were broken up in very energetic collisions. Lead, thallium, gallium, germanium, krypton, tin, iron, arsenic, and many others, have been found in intergalactic clouds, so it follows that there is plenty out there. Heavy elements, though, are said to be made in supernovae; but, as we'll discuss shortly, supernovae don't happen very often, so very long times are required to make up all of this heavy material.

More Antimatter:

Yes, antimatter is real enough. It is widely used in particle accelerators and medical and physics research. A someday projection is that it may be used as a super-efficient fuel! Every existing particle is said to have its antiparticle; like a mirror image. Not only is it the opposite of matter as we know it, it is the *natural* opposite. And, opposites do attract, but when matter and antimatter collide, they are completely transformed into gamma-rays or light.

The Big Bang scenario requires a billion times more matter and antimatter than all the mass that exists in the universe today, ...to be born together and destroyed together, ...instantly, ...and that all of the remaining mass that we observe today is supposed to be matter, and only matter. Never mind the wastefulness – the normal, ordinary, natural application of forces and particular polarities of energies must preclude the simultaneous creation of two opposites, heat and cold, light and dark, positive and negative, matter and antimatter, or any other such contradictions – in the same spot. It follow then, that if we can argue that all the antimatter in the universe was *not* destroyed in that Big Bang, then we might question the idea that everything that we see in the universe is only matter. ...And we will.

Gravity:

Gravity has been raised to the most majestic of forces, partly, perhaps, because relativity has given it such significance; partly, maybe, because most scientists have not considered other forces as instrumental in the universe. In just about any system, gravity, magnetism, and electromagnetism play a role, but in the laboratory, gravity is the least meaningful, the very weakest. We have been sold on the idea that gravity is the dominant force in the universe, but electromagnetic forces exceed gravity by *40 orders of magnitude*! Gravity is recognized as probably too weak to form the structures and arrangements of galaxies that are observed. Gravity is so weak, that in spite of the most careful attempts, in spite of the marvelous advances in measurement technology, researchers can't pin down the magnitude of the gravitational constant's value.[59] In spite of the fact that you and I might be too heavy, the force of gravity is weaker than our arguments here today. Some physicists, trying to make sense out of a frustrating gravitational universe, have even proposed letting the gravitational constant vary with time![17]

Gravity waves are important predictions of general relativity, but despite Herculean efforts and sensitive detectors, still today, no detector can find them and no hope is forecast. No gravity waves have been spotted. If gravity waves exist, there must have been one humongous gravity wave at the Big Bang, but there seems no evidence of it. One reason why they might not have been detected could be that gravity, in spite of theory, might have a limited range!

Gravitons are described in the Definitions Appendix as hypothetical particles that carry the gravitational force between masses. They are supposed to travel back and forth freely between the coupled bodies, and, of course, would have to act similarly with either body, even if one of those bodies was of invisible, exotic, proposed dark matter. Gravitons would have to travel at much faster than the speed of light, because their effect over long distances can be measured as instantaneous. Big Bang believers might have to agree with that, because the Inflated universe is said to have greatly exceeded the speed of light, and it certainly couldn't have traveled faster than the gravitons, or we wouldn't have gravity today.

Gravity has been forever the "mysterious" force between objects, and Einstein's offer of a simpler concept that mass merely distorts space hasn't made things any clearer. How, why, does mass distort space? Perhaps a new theory of gravity is called for. One suggestion is that gravity may not work in the very faintest environs. One has it that gravitons might cause the force of gravity to vary in ways that become noticeable only on vast intergalactic distances.[71] And one, perhaps the best, explains why gravity *must* have a maximum range.

Gravitational Clumping:

Gravity was to have become a "distinct force," almost instantly, at about 10^{-32} seconds into the Big Bang scenario. Perhaps you remember, ...that the gravitational force between two masses increases directly with the size of the masses, but that it *decreases* with the square of the distance between them.

$$\text{Gravity} \quad F = G \ \frac{mm^1}{d^2}$$

Well now, even though scientists can't determine the true value of G, the equation clearly shows us that the gravitational pull between any and all particles of mass could *never* be as strong again as it was in that first instant when everything in the universe was compressed into the size of a raison. No matter how great was the explosive force trying to scatter everything, the radial separation of two parts of a sphere increases only directly with distance, while the gravitational attraction between them *decreases* with the *square* of that distance; and if those two parts were to ever get together, they had better do it ...right now!

Any and all particles experiencing radial separation, ...such as in an explosion ... of a hand grenade, perhaps ...whether here or in space, are doomed to an ever-decreasing mutual attraction. Period.

But, if it is decreed that these particles are somehow to get together, and even if we make our own guess for the time of galaxy formation, shouldn't we ask why the universe is supposed to have gone from a super-super density; down past the density of stars, down past the density of the air in this room, down to one billion-billion-billionth the density of water before it decided to form galaxies? Was it not until the attraction between particles was the *weakest* before they started to get together? In the raison pudding recipe, at what stage should gravity start adding raisons?

Stars:

Star Age:

Star ages are notoriously hard to determine; they might be as old as 19 or 20 billion years or more. Intermediate mass stars, 3 to 8 times the size of ours, may last for just a million years. Blue: are they young or stripped old red giants? Red: are they old, inflated stars in the dying process, or are they about to detonate as supernovae? If there are valid questions about the formation of stars, or how their interiors function, it follows that estimates of age are problematical at best, evidenced by the fact that a lot of stars are estimated to be twice as old as the universe itself. As was said, star age and other characteristic estimates are based largely on the star's position on a reference system that is called "the main sequence," that statistical rendering of star development, which provides excellent comparisons, ...but no definitive values.

The farthest stars, about 14 billion light-years away, are thought to be the youngest, but how did they get there? Defenders of older universes poo-poo the recent measurements to Virgo that translate into an 8 billion year old universe, citing new cepheid distance measurements that proclaim the universe to be of 20 billion years. But they are accused of ignoring the fact that their selected yardstick supernovae may not be of average intrinsic brightness; that such galaxies are thought to contain Ia supernovae with "higher intrinsic brightness."

By the way, after a supernova explosion, and all the bits and pieces are flying out into space, they are supposed to gather gravitationally to form the newer stars, et al. But, as each bit flew out, it too was subject to radial separation, so the likelihood of its ever getting together with its sibling bits to form new stars is far less likely than its chance of being grabbed by *already existing gravitational centers; the old stars that are already formed.* This means that a continuous accretion of bits in a star's furnace probably adds to its lifetime, and if this is so, the theorized life span (say age) of the older stars, perhaps 18 billion or so, may have no meaning. Perhaps it could be extended to 25 or more. Or more, on and on. And if those old stars keep at it, they might grow up to supernova size and spread the bits around again, and again, and again ...forever. It is quite plausible, even likely, perhaps mandatory, ...that the number of stars in any one garden of galaxies remains about the same, forever, and the ages of any of them are questionable.

Star Birth:

Gravity and the gravitational clumping of huge hydrogen clouds are generally thought to be the first steps in the established star, or even galaxy, making mechanism, which has seemingly been well understood for quite some time. But even here, in a process almost completely taken for granted as a given, there are unknowns. It is thought that stars can form alone or in huge clusters, or in families of a few dozen or so, perhaps only to drift away. Stars are thought to result from the collision of gas clouds, form in the wispy tendrils of an active galaxy, or deep within its bowels; perhaps all of the above. Uncertainties are a natural aspect of science, of course, but subsequent theories and concepts usually forget or dismiss the inherent uncertainties in the basic "knowledge" upon which more complex scenarios depend.

Cosmologists have identified a very young star on the fringe of our galaxy. This surprised those cosmologists who thought that such a remote part of the galaxy couldn't harbor young stars, because there is very little of the gas and dust out there that is needed to form them, but a stellar nursery is in evidence at the extreme edge of the Milky Way. The relationship of bright hot stars to gas shows that stars are forming continuously. Stellar nursery? Forming continuously? How can continuous creation be taking place if the stimulus for making stars in a Big Bang has been expended, ...and there is no shrinking cloud in the neighborhood? And, ...we have been told over and again that stars are formed in shrinking gas clouds.

But, the idea that a huge gas cloud could be subject to so much self-gravity that its ambient gas pressure is ultimately overcome, just doesn't square. In the beginning, that cloud is at a density much, much lower than anything we can create in modern laboratories, almost at the density of outer space; at that billion-billion-billionth the density of water mentioned a moment ago. A "huge" gas cloud was just that: *huge*. The mass of that hydrogen cloud had to contain all of the material that would ultimately make up the stars, the clusters of stars, and all of the stuff that would become the galaxy, if that's what is in the making. A cloud with all of that mass, at that extremely low density and pressure, had to extend over such a vast reach of space that the gravitational attraction, that weakest of all of the forces by many, many orders of magnitude, of one side of the cloud for the far side of the cloud, could not possibly overcome the attraction of those "already existing gravitational centers; the old stars," or the natural gas pressure that we demonstrate every time we inflate a balloon. We've all bust our gut trying to blow up balloons, and can feel certain that there is a lot of pressure to overcome for gasses in space to gravitate into lumps.

To complicate the mechanism a bit, it is now observed that "Not only does gas fall inward, but in some stars, at least, vast quantities of gas and dust stream outward" in narrow jets and "giant peanut-shaped bubbles" of gas 100 times more massive than the Sun.[224] The estimated times involved, and the smooth, inexorable shrinking of clouds into stars, then, has to be reconsidered. External forces, of some kind, that can overcome those internal pressures and drive parts of the cloud together, are a must.

A collapsing scenario cloud could, must, develop a spinning central concentration of material that is to be our new star. That is exactly what is promoted. But that spinning mass would have to spin very fast, much faster than our Sun for example, to absorb all of that ambient kinetic energy. The normal, observed spinning rate of your everyday star, which is much slower than demanded by mathematics, is justified by having that body throw off material as it spins. Throwing off material gets rid of excess energy, allowing the body to spin slower. Great! But how does a soon-to-be star throw off material into a surrounding cloud of similarly very, very dense gaseous material that is collapsing onto itself? Nothing could go very far, it would have to be forced back onto the spinning body, if indeed, it could have ever left in the first place. *No, a collapsing, ever denser cloud cannot explain the birth of a celestial body.*

If you had an extraordinarily large gas cylinder: the biggest that you could imagine, under enormous pressure, so that it becomes liquid then solid, and opened that container to empty space, what would you experience? Nothing? If the cylinder is large enough, and if the gas operates in accordance with the prevailing theory of star formation, the self-gravity of the gas would keep it all within that container! But our own experience dictates that the gas that we release would whoosh off into space, never to be seen again. And don't hold on to the bottle!

Also, if that cloud can indeed start shrinking under the force of its own gravity, then the density and pressure has to increase steadily, uniformly, smoothly, inexorably until at the very center of the cloud, the pressure is, and always is, at its highest. Any celestial body that would or could precipitate from that cloud, then, must be one, and only one, regardless of sizes; there could never be a cluster, or a bunch, or even a pair of anything forming from that cloud. Just the one body. There can't be more than one concentration of mass, period. That is the way gravity is observed to work, ...unless. Unless those external forces mentioned a moment ago are able to have their way with the cloud and provide a roiling, fractious turbulence that can develop more than one center of concentration.

Special Early Stars!

Recent observations have detected heavy elements, especially carbon, in gas clouds that seemingly formed very early after the Big Bang. Their remoteness from galaxies, and the generally accepted theme that heavy elements must have formed in stars that went supernova, leads them to the wonderful conclusion that, even though stars almost invariably formed in galaxies throughout the universe's history, *one* generation of stars *must* have formed before galaxies did![223] Those "special" early stars, then, later would cluster into galaxies. The concept envisioned is, that in the earliest moments of the Big Bang, "there were many more small clumps than big ones," favoring the precipitation of many stars instead of galaxies.

Well, maybe so, but. ...If small clumps can precipitate into stars, how would you ever get a large clump for the precipitation of galaxies? Also, inevitably, the concept of wide-ranging forces and tendencies has to contradict this. Whatever pressure or proclivity or force that could result in the precipitation of those first stars, in a Big Bang, would have *been expended* with that first event, thereby minimizing the likelihood of further precipitation of any kind, perhaps forever. And, ...those stars, all the same age, would be subject to the continuing radial separation and decreasing gravitational attraction that existed before they became stars. And, ...and, ...the stars flowing into (?) galaxies are not seen to be all of the same age. If it were argued that the original stars would have been gone by now, and those seen in galaxies are a born-again later generation, then the more central stars would most likely be the newest. Not at all what is observed.

Star Interiors:

There are still some aspects of star interiors that are unexplained. The many pictures of star explosions, jets; clouds of nitrogen or oxygen or sulfur; all hint at a configuration within that is not explained by a straight-forward fusion process. We can pretty much accept that hydrogen clouds are, and perhaps should be, ubiquitous in space, but what about the heavier elements? Oxygen, for example, ...sulfur. How can great clouds of these elements be spread out in space? Yes, we understand that they probably come from exploded stars, but back in the interior of those stars; were the gasses segregated into huge secluded reservoirs so that on detonation they could maintain their separate identities? In spite of the fact that both oxygen and sulfur and others are strongly reactive? If the various elements were in separate layers before the explosion, wouldn't they intermix as they do here? What would separate them? If they were not to intermix within the star before the explosion, wouldn't they do so during it?

Of course there are still many unknowns about star interiors. Scientists have made great and clever efforts to detect the solar neutrinos that are calculated to exist in the Sun, but results are very much lower than expectations. Also, a number of exploded stars look like doorknobs; or the bow that a girl might put in her hair; or a series of rings or jets! ...Not at all what we should expect from the explosion of envisioned concentric, spherical bodies.

Globular Clusters Of Stars:

Globular clusters are collections of stars, sometimes dense. They are normally thought to be 10, 15, perhaps 18 billion years old, but some have been found to be less than one-tenth that age. It was thought that the globular clusters surrounding the Milky Way were formed at the same time as the galaxy, but recent estimates for the cluster ages are higher than the estimated age of the galaxy itself. One guess has it that the globular clusters result from galactic collisions, (galaxies of different ages, of course) but our particular clusters are too evenly distributed, and there is no evidence that the Milky Way has ever suffered a collision.

Do we understand galaxy development? Is the main sequence a reliable reference? Isn't one or the other in error? The processes in star interiors are supposed to convert deuterium into helium, and that, supposedly, is why "there is less deuterium now than there was in the Big Bang;" but one has to estimate the rate of star birth to calculate what the original amounts were, and those estimates are questionable. When theorists measured the amount of deuterium in "ancient" clouds, the amount was higher than expected. Instead of acknowledging that something was wrong with the theory, they decided that there must be much more dark matter than predicted! In a lot of galaxies the globular clusters nearer the center have heavier concentrations of the heavier elements. This implies that there had to have been a supernova, or two, or more, in the early development of those galaxies.

Supernovae:

When the time comes, many moons from now, average stars like our Sun will die relatively peacefully. They might inflate and contract a couple of times, but then they usually, slowly, condense quietly into a white dwarf. But not red giants ten times or more bigger than the Sun; when they go, they go out with a bang – literally. When a red giant's internal pressures are finally overcome by its own gravity, the huge envelope collapses quickly, perhaps leaving a neutron star or black hole, but certainly exploding with the energy of 100 million Suns – in just a couple of days – spreading most of its stellar debris every which way, far out into space. And it is just this supernova spreading of star stuff that our Sun, and similar average stars, *need* as a source for a lot of the raw material, at least the heavier elements, from which they are grown.

How often does this happen? Some scientists believe that only about 100 or 200[218]   (One Source says only 5 – 10!)[33] or so a century happen in the average galaxy, while others feel that they happen about three times that often. Not many are observed, surely; but one scientist thinks that a lot of the supernovae could be directional – in the form of galactic "chimneys"[33] – so we wouldn't see them unless they were aimed our way. The problem with that, ...looking back on the mechanism of supernova development, with the general star-wide spherically collapsing stellar envelopes and all, we might doubt that directional chimneys are much of a factor, especially since few have been "detected" and perhaps none directly observed.

Stars *need* supernovae debris as a source for their heavier elements, because, we are told, that it is only in the higher pressures and temperatures of those huge stars that go supernova that heavy elements, such as silicon, sulfur, metals, etc. can be formed. If so, the heavier components in our average Sun and all of it's zillions of *kin had* to have come from earlier, much earlier, supernovae; even though supernovae are not observed to happen very often. On top of the infrequency problem, a very large portion of these original red giants is said not get spread around at all; it is supposed to remain in very dense, very heavy neutron remnants. And, some, or a lot, of supernovae produce little or no heavy elements: its just so happens, that if these large stars contain a lot of oxygen, they can go supernova before they get a chance to produce them; so not all supernovae are helpful in the production of this vital stuff. Worse still for the distribution of this critical material for newly forming stars in need of supernova seeding, is the realization that most of the material that does become wide-spread is far more apt to be gathered up by the already established gravitationally attractive older stars. Younger stars, then, must not have much in the way of heavy material, perhaps offering another measure of star age.

Well, now, let's think about this a bit:

> Supernovae are not observed very often.
> Not all supernovae make heavier elements.
> A good portion of the original star remains.
> Each explosion spreads the material in all directions, only to have a lot of it absorbed by older stars.

So, it seems that for our Sun, and all of its billions upon billions of siblings, to have been able to snatch up the heavier elements that they have today, there had to have been one heck of a lot of supernovae. But, since they are not observed very often, all of this growing of red giants, collapsing of shells, explosive distribution, and the ultimate seeding of the universe with this vital stuff, had to have been happening for a very, very, very long time.

Galaxies:

Your new baby might look rather average to the rest of us, but to you; she is the most beautiful baby in the entire world, because you see what you want to see; just like we do, just like scientists do. Astronomy and cosmology are intended to be sciences of observation, but if the observer has preconceived notions, or is hoping for a certain type of evidence, almost any observation can be interpreted in the desired way.

In point, a recently observed NGC 7252 is said to be an elliptical galaxy that resulted from the collision of two spirals that are thought to have been rotating in *opposite* directions; yet the resultant is that of a small spiral quietly and smoothly superimposed on a background of a large, seemingly undisturbed gas cloud.[76] A large group of globular clusters, two young, too many, and too smoothly spread out to have formed in the original galaxies, indicates to scientists that the collision itself might have resulted in the birth of these globulars. A few thoughts:

a. The small spiral is almost exactly centered on the gaseous background, indicating that in spite of the relative motions, we coincidentally observed it at just the right time that they are exactly, concentrically aligned. This might be quite all right, but coincidences are suspicious.

b. These two original spirals that were rotating in opposite directions seem to have merged especially smoothly, yet we are told over and again that the collision of two galaxies can be catastrophic. Similar, more gentle collisions have been used to explain extraordinarily large onergy displays. "A star burst galaxy resulting from the collision of two galaxies with an energy greater than 500 billion suns," for example.

c. Globulars are seen in most galaxies, but most galaxies don't seem to have suffered collision, and seldom is a disturbance of any kind credited with globular birth.

d. Globulars are usually thought to be old, which implies a lengthy development, so it would seem that cataclysmic events are not the likely way to generate them.

You can see what you want to see, sometimes at the expense of logic.

Galaxy Age:

How old is our galaxy? It depends on how you measure it. There are a number of different ways of inferring age: colors and brightness, luminosity, abundance of certain isotopes, probable age of the oldest stars, all arguably valid. The oldest stars in the Milky Way are estimated to be about nine billion years old, while the oldest globular clusters in the halo surrounding our galaxy are estimated to be about 18 billion. As we peer farther afield, we see the older stars, even back when the universe was to have been very young. But the stars and galaxies that seem to have been born billions of years ago looks every bit the same as anything closer by. No evolution is noticeable locally.

Because galaxies are said to require amazingly long times to form, scientists seem to have attributed to most of them ages of between 14 and 16 billion years, in spite of the much younger estimates for the universe as a whole. Many galaxies, including our own for instance, have a lot of globular clusters, which are very large groups of stars in a relatively small space, that shine with the weak, reddish light of old age: as much as 16 or 18 billion years. You would think that the parent galaxy then, the Milky Way in our case, would have to be at least that old, even though its official estimate seems to be about 15 billion years old.

A 16 billion year old star or galaxy doesn't have to be 16 billion light-years away, but a galaxy 16 billion light-years away does have to be at least that old, because it would have taken that time for its light to get here. That galaxy 16 billion light-years away, though, *could* have been 16 billion years old when it sent its light beam towards us. (It certainly *looks* that old.) If so, the universe would have to be at least 32 billion years old! Guessing the maximum age of the universe has to be of dubious value at best.

For those galaxies that do seem to be 16 billion light-years away, how long did it take for them to get there? Since the light is coming from where they were 16 billion years ago, is that where the center was? Not necessarily, because we see such light in all directions. Some possibilities:

a. We are at the original center and everything expanded away from us. NO ONE suggests this.
b. We are not at the original center, and we are seeing "only" 16 billion years of the universe, which is actually much bigger, and therefore, much older. But, as we discussed in the raison pudding discussion, redshifts are darn near at the maximum limits now, so the larger redshifts can't be very reliable.
c. We are not at the original center, and what we are seeing out in back, at 16 billion years, is near the original center; and what is out in front, at 16 billion years, is even farther from the original center than we are. So what is out front is actually twice as old, or 32 billion, because it had to travel twice as far, in spite of the fact that it is measured at 16 billion; and if that measurement can't be counted on, what about the others?

You can pick one, but nobody likes any of them, in part because of that latest 8 billion years measurement, and because galaxies supposedly started forming as early as 200 million years after the Big Bang; and even some of those are reputed to have old red stars by then.[132] But it is observed that gas clouds were still forming into galaxies as recently as a few billion years ago. Young globular clusters have been spotted at the center of our galaxy, indicating that star formation within galaxies is an ongoing activity. And, if that is so, then galaxy formation is an ongoing activity as well, (remember how galaxies of different *ages* were to have collided) and the universe is better considered as a continuous "process."

"It looks like galaxies are created at different times and different places."[24]

Galaxy Movement:

Once dark matter was invented to provide the missing, "needed" gravity to explain the much too fast spinning rates and their speed across the cosmos, scientist found themselves sliding down a slippery slope to surrealism, relying on more and more exotic inventions and rationale to explain the misunderstood; to no avail, as we are trying to reason here. Galaxy movement, at once the provocation for dark matter, is perhaps the clearest evidence that there never was such a happening as the Big Bang; that it just couldn't be.

A number of galaxies have companions, and so does the Milky Way, with its Large and Small Magellanic clouds. Why are the Magellanics nearby? Why are they moving towards us? It is possible, even seems likely, that they formed at the same time as our galaxy did, and if so, each of us would have long ago raked in all of the in-between debris, thereby increasing the mass distance and decreasing gravitation between us, right? How can we be attracted to each other in an inexorable radially expanding space? Remember, "not only was the universe expanding, *space itself* was expanding," ... they tell us. "As space expands, more space pops up in between!" Mutual gravitation doesn't seem to justify the movement by itself; something had to have happened to nudge these bodies close enough that gravity could be given a chance to have its way with them.

Galaxies apparently do travel in all direction, even against the flow of expanding space, hence the occasional collision. Interestingly, the stars in galactic collisions almost never hit each other! They are simply too far apart and comparatively too small; those miles-apart dust particles again, and if a few did collide, they certainly wouldn't disturb their parent galaxies. Gas clouds can collide; they can even steal gas from one another, and they frequently put on quite a show. The unlikeness of star collision, though, has not deterred scientists from using galactic collisions to explain even monstrously large energy sources, such as the so-called starburst galaxies.

One of two disks of gas seen in collision by the Mauna Kea telescope is apparently spinning at 2 million miles an hour. It seems that it would take a gravitational source 100 billion times the mass of the Sun to hold it together! Mysterious indeed, but wouldn't you know, a black hole is suspected of being that gravitational source. If it were a black hole, though, it would be 100 times more massive than thought possible; and seemingly disregarded is the question of where the black hole came from. It certainly wasn't sitting around waiting for a couple of galaxies to come by, and no is suggesting that a collision of gas clouds can form a black hole.

I was recently looking at a wonderful picture of a galactic garden. It really was beautiful. Two or three dozen galaxies of all shapes and apparent sizes; ellipticals, spirals, barred; and oriented in every which way. It was enough to fill you with wonder. ...And I wondered, ...*Why*? Why were they different? Why were they of different types, and especially, why were they oriented differently? We are told time and again that the precipitation of galaxies was from a wide-ranging, far-reaching flow of almost perfectly homogeneous energy from a long way off. *Every* galaxy within arms-reach of each other should be a sibling! Perhaps not identical, perhaps not of the same size, but they should surely look a lot alike. They should certainly share the same orientation. They would have inherited the same angular momentum! It is almost physically impossible for two galaxies born next to each other to be rotating on different planes, unless they have been in collision of some sort! *All* of them? And, the orientation of the spin-plane of all galaxies should vary from their neighbor only in degree as a factor of separation! Detective Columbo would have said that it is *impossible* that the galaxies in that picture could have formed from a single event.

Apparently galaxy collisions don't always result in large energy displays, because many elliptical galaxies are hypothesized to be the merging of two spiral galaxies, and there are a number of examples which seem to be the result of rather peaceful merging. Computer simulations seem to support the idea that spirals and disk galaxies can form ellipticals; and why shouldn't they merge smoothly, especially since stars seldom hit? The expectation is that nothing should happen other than a reshaping of each and possibly an eventual assimilation. It is also possible that the two could go on their merry way without really disturbing each other; but if gas clouds are merged, there should be, and we do see, some feisty fireworks.

But! The energy complications of two galaxies in collision are not why I mention this. What I want to know is how galaxies can collide? At all? We keep having to go back to the damned tenacious radial separation considerations of a Big Bang expansion. We can't picture how such galaxies could possibly collide. Any two of them in the same section of Big Bang space would have to be traveling at about the same speed and in the same direction. Supposedly each was formed by collecting all the intermediate material, thereby creating a void between them, and *increasing* their gravitational separation rather than the opposite. They say space expands. Why doesn't that *prevent* material from getting together? ...How can galaxies collide?

So, again! A universe that allows lateral shock waves and streaming forces from ...somewhere else, is far more apt to arrange the collision of two galaxies. Remember that lone, young star out on the far reaches of our galaxy that we mentioned while we were discussing age? The one that shouldn't be there? Computer studies of the "collision" of more than one body, and the newer laws of chaos that are being formulated, show that stars can be thrown off in any and all directions as a result of such encounters. And so can galaxies go freely about their every merry way in a non-radial expanding, non-Big Bang universe.

But, wayward stars or the collision of two galaxies is trivial indeed, compared to the larger-scale movement of galaxies throughout the universe. Although the accuracy of any of these measurements can be questioned – for lack of an absolute reference – it was determined, in 1986, that a "huge chunk of the universe, at least one billion light-years across," *including the Milky Way*, is apparently rushing toward the Virgo constellation at a furious rate, as if pulled by a gravitation source the equivalent of 10 to 50 billion, billion times the mass of the Sun. This monstrous gravitational source is called *The Great Attractor*, which is, itself, measured as moving towards something ten times bigger! The streaming is measured at a speed equivalent to 1.56 million miles per hour![218] And there are other, different motions that have been discovered. Our galaxy, for instance, seems to move at up to twice that speed.

More than this, a continuous streaming across the universe has been observed which suggests the presence of other extraordinary gravitational sources – masses that make even the great attractor seem puny in comparison – at *opposite sides of the universe, where the density is said to be twice what it is elsewhere!* [What happened to our homogeneity of density?] Astronomers have discovered that almost everything in the universe is moving rather rapidly in "rivers" across the observable sky. They tell us that the expanding universe can be likened to an inflating balloon, but not explained is how any one point on the surface of a balloon can collide with another, or "stream" in unison with its neighbors across the surface. And what about streaming in a raison pudding? Wouldn't that indicate an unevenly heated pot? Not what you might expect from a Big Bang model, but it fits right in with a concept of an infinitely large universe with areas of different densities and forces. The "lack-of-reference" business mentioned a moment ago relates to evidence that suggests that the background microwave radiation cannot be used as a reference for making such measurements, and, as was said earlier, "...maybe nothing is moving at all." [68]

When the large-scale movement of galaxies toward Virgo was detected after very careful measurements for over a year, rather than acknowledging that the evidence raised questions about the Big Bang concept, dogmatists insisted that there must have been a mistake. "It must be incorrect because it doesn't fit into existing theories." A certain amount of hesitancy to fall into line behind maverick ideas is understandable, but what is happening in cosmology is a wide-spread, "official" policy of putting down any ideas that conflict with the party line. Even though defenses for that line are becoming more bizarre all the time; even though the estimates for universe age are contradictory to the extreme.

Structures:

One of the most nagging of doubts has to do with observations that galaxies seem to be arranged in "structures." Scientists using redshifting to plot three-dimensional maps of the cosmos have discovered great structures of galaxies throughout the universe: "great walls" of galaxies millions of light-years thick, hundred's of millions of light-years wide, and billions of light-years long. At the speed galaxies move, some of the structures would have taken 100 billion years to form. The galaxies seem to form a web around bubbles of void that are perhaps hundreds of light-years across, and seem to clump together every 400 million light-years, or so.

Some physicists think that there wasn't enough time for these structures to form it we rely on visible matter alone, that the structures might be at least 63 billion years old! Some have said that they might be *ten times older* than the Big Bang claims for the entire universe! It is suggested, hopefully, inevitably, that invisible, exotic, cold dark matter may have formed first and helped pull things together in the allotted time, but measurements by an infrared satellite seem to discredit the cold-dark matter mechanism of galaxy formation. Big Bangs just aren't supposed to produce such structures.

"There is more structure on large-scales than is predicted by standard dark matter theories of galaxy formation."[46]

"There is no theory using conventional physics that can explain these structures without causing other inconsistencies."[19]

"The periodic structures are like periodic extinctions for geologists. Why should the process that made galaxies pick out that pattern? It is so far beyond our understanding that theorists dismiss them for the time being – hoping that they are an illusion."[42]

"As of now, there is no model that explains why the universe should be the way that it is."[12]

"...there are more structures than predicted by cold dark matter."[12]

Of course, it's possible that the "structures" are merely the result of inaccurate redshift measurements, but even if they are indeed there, they would be much more likely to have formed in an environment of ongoing turbulence then in the sudden whoosh of a one-event Big Bang. Zillions of galaxies are seen to lie on thin, curving sheets surrounding humongous voids, clearly challenging the dogma that the universe is homogeneous, and the long times estimated to construct such structures should have put the lid on the Big Bang's coffin long ago.

Chapter 11

Galaxy Formation

---

* Collapsing Clouds * The Wrong Way * The Right Way *

Because each newer observation of ever more redshifted galaxies keeps placing mature galaxies, some containing old red stars, closer and closer to the big event, scenario devotees are forced to lessen the required time for forming them. Recent estimates are that galaxies started forming as early as 200 million years after the event. From a rushing sea of particles trying to race away from each other, "chance fluctuations" are to have created tiny lumps of gas, and these denser regions having more gravity than the others, are to have attracted matter from the sparser volumes, in spite of the gravitational equation that shows mutual gravity decreasing with each moment; each yard traveled. Two hundred million years isn't much when you consider that the Earth is estimated to be about 4.5 billion years old.

If you picture a cloud of atoms streaming in the one specific Big Bang direction, out, you can imagine the difficulties they would have in trying to join. If you could ride along with one of those atoms to observe the group a little better, you might experience something like we do on an expressway; all the cars are going fast, in close formation, but with little or no motion relative to each other. The cars don't want to come together of course, but neither do the atoms. The progressive radial separation, conjectured smoothness, and certainly the natural repulsion of similarly charged particles, is simply too much for the decreasing gravitational attraction, as we have already noted. All lateral forces are mutually canceled, and the unswerving momentum of each particle is ...straight out. The dilemma of large-scale homogeneity versus small-scale inhomogeneity is irrelevant. Those atoms just don't want to get together.

If, however, we can muster up a very large cloud of particles that are *not* experiencing radial separation, we might have an entirely different story. Just as with the familiar air pressures and tire pressures that we live with on Earth, all particle clouds in space have an internal gas pressure that has to be overcome if it is going to condense. The Earth's atmosphere certainly hasn't condensed very much over the last four billion years, nor has Jupiter's in spite of a mass 300 times greater. It turns out that only very large clouds have enough mass and supposed gravitational self-attraction to permit themselves to begin the shrinking process, a very doubtful situation considering that the larger the cloud, the farther away the mass that is supposed to add to the overall gravitational pull. But, if our very large cloud of relatively motionless particles experienced forces or pressures from without, say from shock waves or flowing streams of particles from ...somewhere else, then our undecided cloud might find itself starting to contract whether it wants to or not. And those shock waves or flowing streams are more apt to be found in a universe that moves in more than one direction. While a very large gas cloud in a radially expanding Big Bang situation would find it very difficult or even impossible to come together, a very large cloud not subject to those radial forces; one that is being squeezed from perhaps more than one direction, would find it very difficult not to. It turns out that there are many, many examples of clouds in furious collision, from which stars and galaxies can readily be seen precipitating in large numbers. "High-velocity clouds have been a mystery for 40 years."

The halo of globular clusters around the Milky Way that I mentioned earlier, doesn't look like an angel's halo, so much as it does a fidgety bunch of seagulls circling for a handout, not necessarily in any pattern. They are mostly old, perhaps heavier, and said to have been cast off as the accreting galactic embryo spun up with the energy it took from the original shrinking cloud. But, even here, there are points to discuss.

It is argued that the globulars were thrown out of the galaxy-proper *because* they are heavier, but that argument would have *all* of the heavier stars out at distances proportional to their weight. Disregarded is the realization that centrifugal force is directly proportional to mass and inversely proportional to distance, and a twice as heavy star in that shrinking, spinning cloud at twice the distance would experience the same force as our other star. And, ...gravitational forces for any particular distance would be twice as great for that larger star. The net result is that the positions of globular clusters seems unrelated to their weight.

The globulars, at least as old as the galaxy itself so they had to form together, are not disposed on the galaxy's plane, but are spread out every which way, helter-skelter. Whereas the stars in the Milky Way, and the planets around the Sun, are almost entirely on a plane of rotation such that a spinning effect seems obvious; this isn't so for our randomly sprinkled globulars, and another mechanism must be considered.

The oldest stars in a galaxy seem to have their highest concentrations at the center, and younger stars are primarily found out in the arms, "which are rich in hydrogen and dust."[218] Fine. The shrinking gas cloud scenario ignores our earlier discussion of "Star Birth" in the last chapter, of course, but if such a mechanism can apply to the making of galaxies, the procedure demands that they must essentially be formed in one "process," for the most part all at once, and there should be only an incremental difference in star ages. The central stars might indeed be the oldest, but those slightly further out must be proportionally only slightly younger, even to those on the outer edges. If age differs only proportionally with distance, (position) the makeup of stars, also, should differ only proportionally, and neighbors should be essentially identical. How red giants destined for the forming of supernovae could develop among quieter types is unexplained in this scenario. Of course, if older is related to heavier, the concept of throwing out the heavier (along with the globulars) means that the older should *not* be near the center, and a contradiction arises.

And, if you remember, two types of dark matter are envisioned, one of which surrounds the whole galaxy. But, how a "surrounding type" of dark matter cloud, which is supposed to come first to help matter gather, could participate in the collapsing of a much larger gas cloud, is simply not discussed. Does the "surrounding" type initially surround the huge gas cloud? Does it simply wait until the cloud condenses? The concept is merely another contrived bit of silliness, but if such a surrounding dark matter cloud existed, it would tend to *negate* the gravitational development of the galaxy, because it would attract matter *away* from the center.

If stars are formed in the "compression" process of a collapsing gas cloud, there should be no dust or light material left in the vicinity. Why, then, is all of that light material out in the arms? And, what force is continuing to bunch the remainder into isolated new stars?

To overcome this consternation, "odd" gravitational disturbances: a continuing series of density waves or shock waves is envisioned, in spite of the purely fictional nature of such an invention. If, an inward flow is being experienced in galaxies instead of an outward one, centrifugal force would insist that it be a relatively gentle movement towards the center, much like a smooth-flowing river, and local gravity would not allow the formation of new stars! The already formed old stars would have far more gravitational attraction than intervening clouds of star-source material, and they would simply keep on sucking up any leftovers. If everything were flowing inward, whatever debris was in the area of the center would most likely become the oldest stars, alright, but precious little material, and certainly not much in the way of pressures or stimulation, would be left for the development of newer stars.

How would a star in space notice whether or not it is being gravitationally effected by another body or group of bodies? You and I experience weight, and can watch things fall. But a point in space, unconnected to another, ordinarily experiences nothing, nothing at all. A moon and planet relationship is different because of proximity, and the fact that one side of the body experiences a different gravity than the other side. But not so our point in space. It would have to get very close to others before it could notice anything, anything at all. Age, size, material substance, or any other characteristic of the different bodies in a galaxy, then, is not influenced by gravity, in a predominantly gravitational universe; so much as it is influenced by *timing*. *When* the star was formed, is the crucial determinate of its characteristics. Even *where* the star was born is not the issue, because the seed material in the area of the star's delivery is determined by the when-of-it-all also, and one side of that star is not influenced differently than the other side of that star; not by the far away galactic center, it's not. The different characteristics of a galactic body's stars then, ordinarily cannot be determined by any gravity other than their own!

We said that the Milky Way's age is estimated at 15 billion. The oldest stars in the galaxy proper are estimated to be about nine billion. Those stars, about half of the galaxy's age, would have to be second or third generation. This continuity implies that there are bound to be fourth, fifth, and further future generations. It could also imply that those nine billion year old stars *could be* fourth, fifth, -- or fortieth -- generation stars, maybe, and that the galaxy is far older than estimated. While the minimum age of a galaxy can be scientifically estimated, the maximum age can only guessed at within the constraints of a preconceived model of galaxy formation. The oldest globular clusters in the halo surrounding our galaxy are estimated to be about 18 billion. It is unlikely that the halo clusters formed and hung around for three billion years waiting for our galaxy to form, but the inconsistencies of the Big Bang scenario have forced scientists to concoct many, many such bits of odd reasoning. Most scientists, especially those that endorse a Big Bang environment, have not been able to envision a workable model that explains galaxy formation. Not yet, they haven't.[159]

Have you ever wondered why the sky is dark at night and not completely filled with starlight? Heinrich Obler argued that if stars are evenly distributed, we should see light everywhere we look. That we don't is known as Obler's Paradox. One reasonable argument might be that light from far enough away, in a very old universe, should be too curved, too expended, too depleted by the stuff of space to ever reach here. Another view is that the universe isn't old enough for most of that light to get here yet!

But, still another consideration is the fact that with each new more sensitive observation, billions upon billions of more stars and galaxies *are* observed. One scientist noted that no matter how much more sensitivity he employed, every new image had more faint hints of more; more dimmer, more distant; *more* celestial bodies.[35] The European Southern Observatory, using a high-tech detector, has seen enough dim stars to fill almost the entire sky. No matter how sensitive the film, there is always faint lights hinting at more to come. Billions more. This indicates that redshift is no factor in Olber's paradox.

As continuously improving technology and ever-sharper imaging permit, more and more dim bodies are indeed detected. We should expect this sort of thing as an on-going, never-ending experience, now and forever, but scientists seem surprised each time additional dim bodies are detected; and invariably they are treated with significance. Dim clouds being detected lately are said to be about 20 times more massive than the Milky Way, but 10 times greater in diameter. (Of course, if they aren't as distant as thought, they probably aren't as big as that.) There is supposed to be as many of them if not more, as there is of *all* the more familiar, brighter galaxies that we have seen all of these years. One survey of very dim and distant galaxies resulted in

numbers (and density) much too high for the outer volumes of an
expanding universe. They are said to be huge disks of hydrogen, large
enough to outweigh normal galaxies, but not compact enough to form
stars in large numbers. Some astronomers have suggested that these
so-called dim galaxies may be expended remains of normal galaxies
that were once bright, but others point out that some of them contain
stars that are hot and blue, indicating that they are young and just now
forming stars.

So, here we have yet another age problem. Are they young and just
forming, or are they old and dying? If they are just forming, then we
are experiencing an ongoing birthing, a renewal, a continuation that can
only be attributed to an ongoing universe. If they are old and dying,
then the universe has to be much, much older than the 15 or 20 billion
years that is attributed to the normal, healthy, mature galaxy that we are
used to. If you take the 15 billion years attributed to our dear old Milky
Way, which is still in the star-making business, still in fine fettle, still
full of life and a long way from becoming an old and dying dim cloud,
then we must estimate that it will take many more billions of years
before it reaches the stage – the age – of those "expended remains of
normal galaxies that were once bright."

Not all of the galaxies in a cluster are the same age; sometimes
there is a considerable difference. Compact groups of galaxies, usually
about four or five in dense formation, not part of a recognized cluster,
frequently have discordant redshifts. Galaxy centers are frequently
observed to be shooting out jets and magnetic material, and the activity
is invariably attributed to black holes. (The very scenario of black
holes spewing anything is incredulous.) It is obvious, and scientists
agree, that the concept of galaxy formation is not at all understood, but
the reason might be that pre-conceived notions of Big Bangism and the
preeminence of gravity's role in all of this has obscured a clearer
vision.

We have large ellipticals, spirals, monster galaxies and dwarfs.
Large ellipticals tend to have more globular clusters. Ellipticals *are
said to be* the earliest, immature form of galaxy or the result of the
collision of two older spirals. It is estimated that about 75% of all
galaxies are spirals, *said to be* the mature form, and the spiral arms are
attributed to pressure waves: "odd" density waves, they are called.

But they have it just *backwards*! Cosmologists admit that they
don't understand how spiral galaxies can form, and that might be
because they simply *ignore* the involvement of other forces; they
continue to *assume* that gravity shrinks clouds into galaxies; and they

assume that spirals are the older, more mature form.   But, **spiral galaxies surely have to be the *youngest* type.** Galaxies *have to* be formed in the throws of an electromagnetic birth; from the extraordinarily high amounts of energy acknowledged to fill all of space; to only then decay into elliptical clouds of non-description as the much stronger, but temporary, electromagnetic forces die off. If so, ***and it must be***, the non-planar distribution of globular clusters makes more sense.   Globular clusters are usually thought of as some of the oldest bodies in the universe, and as we just said, "large ellipticals tend to have more globular clusters." Why? Because the ellipticals are the decayed remnants of spirals! The globulars hung around because they are older! If sprightly, crisp, new-looking spirals *are* the younger types, **and it must be,** we don't need "odd" gravitational disturbances, or "odd" continuing density waves, or all of that "needed" dark matter. The older globular clusters far from galactic centers, and that otherwise mysterious "stellar nursery" in evidence at the extreme edge of the Milky Way also make more sense. The younger spirals: the 75% of all galaxies then, represent recent birth; on-going, never-ending birth; in a vibrant, never-ending, universe.

It may seem strange at first, but the electromagnetic nature of galaxies is actually very well established. (See the Plasma Universe in Chapter 20, and Quasars in Chapters 12 and 21.) A galaxy is *an electrical thing*, equated to a dynamo with its arms like armature windings. Magnetic fields in the arms of such a dynamo could more understandably explain the observed inter-dispersed concentrations of dense dust and gas, ...and iron. Might that help explain why new stars have iron? *Are* they spinning? Sure, but what we are probably measuring is the circulating plasma flow of a temporarily expanding solonoid-like event. Certainly no black hole is required in this type of galaxy. Surely we can develop a galaxy, or a universe, without relying on exotica.

Chapter 12

The Biggies

---

* Quasars: Birth? Powerful? Sources? * Neutron Stars *
Black Holes: Exist? Origins? Don't Last *

Quasars:

Quasars are "unknown mechanisms;" "improbable monsters." They emit an extraordinary amount of light, radio signals, infrared cosmic rays, x-rays, and, "huge rivers of gamma-rays." The "disturbed core" of galaxies; some seem to be moving away at 90% the speed of light. "The most brilliant objects in the universe, brighter than 10^{15} Suns." (That's a *lot* brighter!)

The most mysterious and powerful objects in all the universe seems to lie at the heart of some galaxies. The quasar is extraordinary to say the least. Thought to be no bigger than our solar system (Their light can change in just a few days, so this is taken as good indictor of their small size.) they are thought to emit more energy than 1,000 galaxies of 200 million, billion Suns each!

Astronomers tell us that they have discovered the most distant quasar in the universe, one with an extraordinary redshift of 5.74, about 12.22 billion light-years away,[290] but such records are broken every once in a while. Also, ...the ROSAT x-ray satellite discovered dozens of previously unknown quasars about 10 billion light-years from Earth, in a region dense with clusters of quasars, and because these distant quasars have about the same density of those much closer, the universe would have had to begin clumping and making quasars earlier than theories allow; indicating that the early universe was just as clumpy as today's. This all means that most of these quasars would necessarily have been born about one billion years after the Big Bang! Most scientists think that quasar energy is too enormous to have come from normal stellar nuclear fusion, and (of course) that it must be the burst of energy escaping from a black hole[170] as it gobbles up gas and stars near the center of galaxies.

But, how could a quasar, that is supposed to be fired by a black hole, be born so soon after the Big Bang, when a black hole, would surely have to take much, much longer than even a galaxy does to form – if black holes are developed from a steady accumulation of mass, that is. Galaxies are thought to be about 14 to 16 billion years old, and at their centers are quasars with the energy of 1000 galaxies. So, how long does it take to make a quasar? Fourteen billion? Shouldn't anything that powerful and complex take longer to form than galaxies? Which came first? And that black hole at the center of the quasar! Whether black holes are the accumulation of mass, or the residue of supernovae explosions, how long does it take to make them? If one forms before the other, just how long does all this take – in a universe that might be only 8 billion years old?

Scientists have determined that quasars are so powerful because of the remarkable energy that they display considering their distance. But, is that distance correct? One has been detected in Cygnus A, a galaxy "just 600 million light-years" away. One scientist, at least, believes that only nearby sources can give off detectable x-rays, so quasars would have to be closer-by (and younger) than thought. Remember Virgo.

Also, quasars are *believed* to be a non-thermal light source, making them unlike anything else in the universe, and it may be impossible to measure their distances with any certainty due to "coherence" considerations of redshift. It was discovered that certain type of light sources, and quasars may be of that type, actually change their redshift with distance because of a coherent relationship of their zillions of individual microscopic radiators. Suffice to say that quasars are a different breed of cat; scientists seem to have no idea how far away they really are, or how powerful, or what causes their energy display; which may not be as bright as we think if we can't measure their distance.

There are detected, galaxies in seeming physical connection with quasars. Although this does not constitute evidence, it could lead us to think that one could possibly spawn the other. Quasars are said to have a propensity towards clustering around galaxy groups, suggesting that there might be a relationship. Latest determinations are that even ordinary galaxies, like ours, spew a lot of x-rays; grist-to-the-mill, of course, for those that believe black holes are in most galaxies; but quasars generate a broad-band of energy as well as x-rays, equaling hundreds of thousands of time more energy than the galaxies themselves. Quasars have been sometimes suspected of *feeding* on a host galaxy for its energy. Another, more palatable suggestion has it that galaxies might be parent bodies for quasars. ...Or the reverse? Indeed, ...*or the reverse*!

The above reflects on the prevailing thought process, demonstrating how myopia distorts the big picture. *Think, for a moment.* "unknown mechanisms, improbable monsters, huge rivers of x-rays and gamma-rays, small but more brilliant than 10^{15} Suns, usually associated with galaxies,...even in the early universe?" Isn't there a correlation with the just proposed concept of galaxies being *sources* of energy and material? We will talk about quasars again.

Neutron Stars:

Neutron stars are said to be the sometimes remnants of super-giant stars that have exploded as supernovae. Their mass is similar to that of our Sun, but they are much smaller; ordinarily not much more than 10 or 12 miles across, and super-dense to the point that protons and electrons crunch together to form neutrons; one teaspoon possibly weighing 50 billion tons! (A white dwarf star has about the same mass, but it is several thousand times bigger.) Neutron stars are said to be highly magnetized, and are found in compact x-ray sources and gamma-ray busters; some spin so rapidly that their emissions are seen to "pulse," and those are called pulsars.

It is said that if the original star from which a neutron star formed had very much more mass, it could have resulted in a black hole instead! Most supernovae are supposed to form neutron stars, but after studying well over 100 supernovae in the Milky Way, scientists could find only a few; so, ...it is said that black holes may have been formed in lieu of the missing neutron stars! But, the idea that any star could explode, spewing almost all of its material elsewhere, and still have enough crunched residue to form a black hole, or even a neutron star for that matter, seems worthy of discussion.

Black Holes:

Black holes are mathematical conjectures, unseen, unproved and probably not provable, but very popular and studied in exquisite detail. We have all heard that black holes are gravitational vortexes so strong that no mass or energy can escape their attraction; and once inside the so-called "event horizon," even light cannot get out. Or time! Whatever ventures near is doomed to fall into the pit. There are supposed to be a great many of them out there, even at the center of our own Milky Way. So many are predicted that you'd think that a lot of them would have been detected by now, but there have been precious few examples of what are claimed to be detected black holes, and there have been none positively identified.

Well, in spite of the concept that no mass or energy or light or …time is supposed to be able to escape from a black hole, grand shows of extraordinary energies, usually associated with quasars, are of course, blamed on those very popular black sink-holes. The emissions are deemed to be so great that black holes are the explanation of choice for the larger displays. ("What else could it be?") The great gushing of material kicked back out from "there" has to be another of those terribly inefficient processes, much as was the Big Bang itself; so, …the universe's two most extreme events, the two greatest examples of energy transfer, are its two most inefficient processes. Even if born instantly in a supernova collapse, what with all that energy being wasted, it would take one heck of along time for even an incremental increase in the development of a black hole.

Now, those extraordinary energies can't be ordinary energies, can they? As we said, cosmologists can be in the habit, inadvertently, of ignoring ordinary word meanings sometimes, but "phenomenal" outpourings *are* seen, in many, many places throughout the universe. The displays are reminiscent of nuclear forces, at least, and they certainly cannot be considered ordinary. The scenario given us, implausible as it seems, is that the material not yet captured, still outside of the event horizon, is suffering such disturbance that it radiates profusely; and, of course, the disturbance that the material is said to experience, close to the line of no return, is understood to be due to the very strong gravitational forces.

We pointed out that a body in outer space probably would not notice any gravitational effects, but a body near another gravitational source might well experience a slightly greater attraction on its near side. It is held out to us that in the vicinity of a black hole, that attraction to the near side would be significantly greater. We are to picture matter heating up by friction as it is drawn into a black hole. Why? One explanation is that the forces on the leading edge of a particle so far exceed the forces on the trailing edge that stretching (heating) takes space. Well, just how much heat *can* be generated by stretching? Here on Earth, we can stretch an iron bar until it pulls apart, with only a minor warming of the bar. So, ...don't try to heat your house by stretching the furniture.

The gradient of all gravitational fields is a gradual one, decreasing only with d^2. Since the event horizon would have to be some distance from the black hole, the distance from one side of our "disturbed" body to the other is insignificant in comparison. There cannot be any appreciable internal gravitational effect on a freely falling body – at all – so none of the grand energy shows under discussion can be caused by gravity! Bodies falling into a black hole probably should be approaching the speed of light, individual particle mass should increase, and relativity shrinking should be taking place, which might negate any stretching. But since our measuring devices would experience the same distortions that the mass does, we shouldn't be able to determine that anything was happening! And if it can't be determined, is it? If gravity is not a player here, neither are black holes! So, ...*none* of the "detected' black holes are valid observations!

We are told that many black holes could have formed shortly after the Big Bang, indeed that "they can pop in and out of existence like virtual particles."[92] (If you think of a black hole as a particle, the mathematics will treat it as such.) Well, this may be so, but virtual particles have been proven real, and they are merely atom size and small enough that the concept of their precipitation from an energy field seems more palatable than similar condensation of a black hole with its much, much greater enormous mass. Energy fields, in any dielectric, are smooth, continuous, void of any discontinuities that could precipitate anything but basic particles. Hail stones can develop only gradually from a single drop of water, and only after a number of repetitive events; and a zillions-of-times-larger black hole must also.

If black holes did not precipitate like virtual particles, then all of the long-term time-taking considerations of star, and quasar, and galaxy development should come into play. But, we were recently told that black holes could be formed in lieu of neutron stars as remnants of supernovae, instantly; there would be nothing gradual about this process, because it is the crunch of a monstrous explosion. But supernovae can form only from the earlier development of a red giant and its ultimate collapse, all of which take time. We are told that some or most quasars have black holes, and some or most galaxies have quasars. If black holes are at the center, then it follows that the black hole would have to form after the galaxy formed. Apparently, black holes aren't just lying out there waiting to gobble up passing galaxies; that sort of display would be very noticeable, yet not very many such happenings have ever been noted. So, it's galaxies first, then black holes.

A black hole is supposed to have been confirmed to be at the center of the Milky Way. ("Nothing else could explain all of the energy spewing out.") But now it seems that not enough gamma and x-ray radiation is being detected for it to be a black hole. Wait. Again, rather than admit that earlier evidence for this most exotic of the models could have been premature, another amendment is offered. This *particular* black hole, the one closest to us and most observable, is "different" from the rest, in that it swallows almost all of the radiation that black holes are famous for before it gets a chance to radiate.

Scientists believe that after a "chance fluctuation," most of the stars in a galaxy form together; and a galaxy begins all the eons of cloud shrinking, partitioning, and shrinking again, until at last each section of cloud is too dense to allow radiation; and the stars finally precipitate out in large clusters, pretty much all at once. But, the concept of pervasive wide-scale formation imperatives dictates that the compulsion that caused all those stars to crystallize all together throughout the broad volume of the galaxy *could not* have caused a black hole to form. The localized, extraordinary density required for black hole manufacture would be nothing at all like that of the rest of the system. Since the black hole could not form before – or during – the galaxy's creation, then any black hole observed in a galaxy would had to have developed well after the galaxy matured. But, the significant dark mater argument negates even this effort.

If a black hole was to have been born in a supernova, that supernova had to have exploded in one of the most awesome spectacles of all time, and such events are routinely use to excuse the very largest voids of empty space in the cosmos. While a supernova may have left a black hole at its origin, there should be nothing whatsoever left in the vicinity to further feed the phenomenon! *If that is so, how can we ever credit a black hole with those great, extraordinary emissions of energy?* The billions of bits being swallowed by a black hole ...bit, is hard for us to swallow as concept.

The latest is that black holes are not forever. Scientists estimate that black holes could *evaporate* when the universe gets cold enough:[92] in 10^{69} years! ...10^{69} years! If you line up all of those zeros, they would stretch to Chicago! So, ...it took perhaps as little as 8 billion years for the universe to drop from a billion, billion, billion degrees to 2.73 Kelvins, ...but it is going to take a billion, trillion, trillion, trillion,

trillion years to cool off those last 2.73 degrees! The mathematics is undoubtedly correct, but of no use to people that live less than 10^2 years.

Neutron stars are almost black holes; they, also, are super-dense, super-strong gravitation sources; supposedly they were formed the same way, so why aren't neutron stars credited with similarly prodigious displays of boiling and bubbling gushes of energy generally blamed on black holes? Indeed, why are they hard to spot at all? Why were no neutron stars supposed to have formed in the Big Bang? Why don't they (or normal stars) precipitate out like virtual particles? They would have been easier to form!

Advocates count on mathematics to define black holes, but if it were not for mathematics, *black holes might not exist.* I am not at all questioning the mathematical expertise of those that insist on the existence of black holes, but it is possible that the mathematics do not accurately represent the relationships of particles under compression. Sometimes models only approximate nature. If the mathematics is truly descriptive, there may indeed be such things; but remembering that gravity is one of the weakest forces between particles, and that inter-atom forces are very, very strong, it seems that the odds are very much against them. Some other models of the universe do not require black holes, and one, at least, prohibits them.

"...a sufficiently strong pulse of light can collapse to form a black hole."[86]

Oh, well....

Chapter 13

Inventions and Hypothetical Concepts Required to Support The Big Bang

---

* A list of Imaginings *

Ordinarily, in *all* scientific disciplines, *one* unsupported guesstimation would be enough to invalidate a conviction; but here, almost *every single aspect* of the Big Bang theory is hypothetical, invented, intestable, perhaps never to be detectable. The espoused cosmological dictums are based on mutually supporting cosmological and quantum mechanical mathematical derivatives and conjectures that are continually adjusted and readjusted in efforts to overcome their failure.

The following is an amazing, imaginative, surely embarrassing list of exotic phenomena used to excuse observations that don't back up the theory:

Anti-gravity: Theory requires that in the Big Bang beginning moments, for just the tiniest conceivable portion of a micro-second, a temporary repulsive gravity existed; then, in the very next tiniest portion of a micro-second, positive gravity "formed." Recently, anti-gravity has been "detected" on the fringes of the universe.[269] Not around here, of course; out there.

Axion: Proposed dark matter particle. Billions of times smaller than the proton. No limit to the number that can occupy the same space. Reacts with practically nothing, acts only through the weak force, and can have almost any mass desired.[69,72]

Black Hole: Proposed as an almost certainty, a collapsed star so dense that not even light can escape its powerful gravitational field. Supposedly "seen" in a few galaxies.

Bosons: A class of particles whose angular momentum is a multiple of Planck's Constant. They are the so-called force carriers.

Champs: Charged massive particles.

Cosmic Strings: Proposed infinitely long, extremely dense, (They have a thousand trillion tons of mass for every inch of length.) "Thin seams in the fabric of space." A "stretched" singularity.

Cosmological Constant: A repulsive force that would counteract gravity; independent of matter. Supposedly can be zeroed out by wormholes. If it were very large and positive, space would be distorted. If it were very large and negative, there would be no space. But, all the mathematicians and all their horses cannot get the cosmological constant to approach zero – or even close to it. Supposed to have dominated the expansion of the universe after cold dark matter did its part. Redshifts, by the way, show no evidence for either a positive or negative cosmological constant.

Cryptons: Hypothetical "partner" of ordinary matter.

Dark Matter: Proposed invisible particles that are supposed to have formed sooner then matter, would have formed "pockets" of high gravity and pulled in the later-forming matter, such that these pockets would grow into galaxies that would drift together to form structures. Would dominate the expansion of the universe for the first half of its history before the cosmological constant would take over. Invoked to provide enough gravity to keep galaxies from spinning apart. Two forms: slow moving "cold," and very fast moving "hot."

Domain: A carryover from magnetic domains, it is the volume inside a surface-like defect called a domain wall that forms between regions of different densities and velocities of matter. Supposedly, one encompasses our universe, separating us from other universes, out at about 10^{35} light-years.

Duality: Makes elementary and composite objects interchangeable: whether a particle or other entity is irreducibly fundamental or is itself made up of even more fundamental entities depends on your point of view.[201]

Extra Universe Dimensions: Ten, eleven, or twenty-six, depending on the argument.

False Vacuum: A theorized (bizarre) state of matter with a super-high energy density. Bubbles of false vacuum are supposed to be able to create themselves to fill space, maintain density, and give birth to other universes.

Flat Space: The energy density is just equal to the critical density so that expansion will ultimately settle down to a "forever" universe. This form of space is described by our familiar Euclidean geometry: it's the very space that you and I are familiar with. What makes it hypothetical in the context of this list, is that *the mathematics holds it to be highly improbable.* It is by far the universe of choice, but scientist cannot bring their theories anywhere close to justifying it.

Global Texture: A theory for the seeding of galaxies; results from unified field theories that are more complicated than cosmic strings. The vacuum is supposed to be threaded by Higgs' fields that knot into areas of high energy density and "radiate blast waves" that create the voids seen in the structures.[17]

Grand Unification Theory (GUT): GUT is the attempt to tie the different types of forces, other than gravity, into one basic concept. It has become an all-consuming effort that rests on very shaky ground, even though this and TOE (The Theory of Everything) are quite valid philosophically. Such concepts are deemed critical for the Inflationary Model. There is no known way to test it.

Gravitational Radiation: Hypothetical ultra-weak disturbances of spacetime, normally synonymous with gravity waves. Not yet reliably detected.[270]

Gravitino: The fermion partner of the graviton. There are theories with more than eight types of gravitinos, but they require a universe of eleven dimensions.

Graviton: A quantum-mechanical manifestation of gravity fields. It is supposed to be a boson, or force carrier.

Gravity Waves: Predicted by Einstein but none have been detected.

Great Attractors: Equivalent to 50 billion, billion times the mass of the Sun. Or bigger! Invented to explain otherwise unexplainable galaxy movement.

Higgino: Hypothetical "partner" of ordinary matter.

Higgs Boson: A brilliant bit of mathematical invention; a massive particle, surely undetectable. Invented to account for symmetry breaking in the first instants. Its existence is crucial to the Standard Model, and the search for it was the main argument for the Super-conducting Super-collider (SSC), which congress killed. Higgs bosons are imagined to "suffuse the universe like a dense fog."

Higgs' Field: A very, very specially shaped mathematical "bowl" with a very, very specially shaped slope and bottom depressions that would allow a mathematical marble to move in a extraordinarily complicated way, to include going through (tunneling) the side of the mathematical bowl as it sees fit, from the "wrong vacuum" to the "right vacuum," that is supposed to allow Inflation to work! "Fields thread the cosmos in swirling patterns, that collapse into knots of high energy density, that then melts the Higgs' field, that then unravels, and the energy radiates away in a spherical blast wave."

Hubble Constant: Relates to the rate of the universe's expansion, which results in an estimate of it's age; a high value indicates a young, eight to ten billion year old universe, while a low value suggests an older universe, perhaps of 20 billion years. No one can agree on its value.

Inflation: An artificial variant in the proposed expansion of the universe that negates the initial constraints. No matter how the universe began, inflation was to smooth out the disagreeable characteristics and reinforce those desired. The original Inflation model had "extreme" inhomogeneous problems, but by selecting "very special" parameters, those undesired aspects could be discarded. Invented to allow the "horizon" – the distance light could have traveled – to get out in front of matter. The universe was to have expanded for a short time much faster than the speed of light, then settle down to the rate now observed. (Not that we know what that rate is.) Depends crucially on GUT, not yet achieved.

Inflatron: Needed by Higgs' theories to "make inflation work."

Knot Texture: After inflation, space was supposed to have been infused with "knots" of concentrated energy that emitted blast waves as they gradually unraveled. Supposed to explain how particles were forced together to make galaxies.

Large Neutrinos: Hypothetical "partner" of ordinary mater.

Lightweight Cosmic Strings: Unlike the much heavier version, this type is not involved in structure making, but would provide the missing energy as an alternative to quintessence.[274]

Loitering: The stopping and restarting of the expanding universe. No explanation is given for either.

Machos: Massive Compact Halo Objects. Proposed *invisible small stars* of dark matter about the size of Jupiter! Believed to be scattered throughout the galaxies in abundance, but they still make up only about 20% of the mass needed.

Magnetar: "Unimaginably dense, this star would have a solid crust [of iron] covering an exotic liquid core. More importantly, it would bear huge magnetic fields whose motion would heat up the surface to such an extent that it would crack [and allow bursts of gamma-rays] under the appalling strain."[284] Presented as the explanation for "the most intense burst of gamma-rays ever seen" (in March of 1979).

Microwave Background Radiation: Held out as evidence of Big Bang's earliest moment: the recent name for what used to be called the natural "radiation of starlight."

Monopoles: We have north and south magnetic poles; monopoles are one *or* the other, supposedly existing on their own. Supposedly 10^{16} times heavier than a proton, not discovered. The ends of "semilocal strings" are supposed to be studded with them.

Neutralino: Proposed dark matter particle; billions of times more massive than an axion. Travels at 200 miles-per-second cruising speed. Undetected.[71]

Odd Gravitational Disturbances: A proposed continuation of effects that might explain the shape of spiral galaxies, and how newer stars can be born in the galactic arms that otherwise couldn't produce them.

Omega: The ratio of the true average mass density of the universe to the critical density needed to just barely close the cosmos.

Other Universes: To say that it came from another universe is like saying that it came from heaven.

Photinos: Hypothetical "partner" of ordinary matter.

Pinch of Nothing: That from which the Big Bang gave us the universe.

Probability Waves: That which is the recent description of matter.

Quintessence: Proposed new form of energy that adds a cosmic density that *varies* with time and space. Invoked because even all of the envisioned dark matter was not providing the density required by the Big Bang.[274]

Semilogical Strings: Introduced to help the transition from smooth to lumpy because cosmic strings match "neither the distribution of galactic clusters nor the patchiness of" the microwave background radiation. Formed with cosmic strings but ends have magnetic monopoles.[287]

Soft Boson: A type of dark matter particle that has a very tiny mass, but they are *bigger than galaxies*; they can overlap with each other, but they are too big to fit into galaxies. Explains why we don't experience dark matter here on Earth.

Solitions: Clumps of string fields thought of as types of black holes.

Space Didn't Exist Before: Nothing did.

Spacetime: As something mystical.

Spacetime: Pinch of.

Spacetime Defects: Domain walls, cosmic strings, monopoles, global textures, …seed for galaxy formations.

Spacetime Fluctuations: In the geometry of.

Speed of Light Exceeded: For the rest of eternity we have supposedly experienced the speed of light as a limitation, but in those "special" beginning Big Bang moments, a temporary permit for mass to greatly exceed the speed of light was issued.

Spinars: Have a mass about 10 million times the Sun but tiny. Can form black holes.

Super-gravity: An "adjustment" to the theory of gravity.

Super-symmetry: An "adjustment" to the theory of symmetry.

Tachyons: Neither matter nor antimatter, but hypothetical entities that by definition perpetually move faster than light.

Temperatures of 10^{27}: Normal star interiors throughout the universe and throughout all time, are thought to reach temperatures of near a billion degrees Kelvin, and every single element, other than hydrogen and helium perhaps, is thought to be formed in these conditions. But the Big Bang, in just that one spot, is imagined to have reached temperatures of a billion, billion, billion, billion degrees temperature, 10^{27} K, for just that one time – that very first second, and only that very first second – out of all of the eons and ages for the rest of all eternity.

Texture: After inflation we were left with a knotty texture to space, and when the knots unraveled, galaxies resulted.[217] It can ripple, distort.

Theory of Everything (TOE): An attempt to include gravity in the Grand Unification Theory. No known (or envisioned) test. Einstein is said to have wasted the last 40 years of his life looking for it.

Vacuums: False, right, and wrong.[132]

WIMPS: Weakly interacting massive particles of dark matter. Proposed massive particles that act only through the weak force.

Winos: Another hypothetical "partner" of ordinary matter.

Wormhole: A bizarre, small pinch of spacetime; tinier than a particle. The mouth of a cosmic string "tunnel;" one mouth might or might not attach to a point in our universe; and the other mouth might or might not attach to a point in another universe. Might exist at the core of a black hole.

Zinos: And, yes, one more hypothetical "partner" of ordinary matter.

Well. There are others, certainly. And they keep coming. Each new realization that something is wrong results in another conceived corrective concept, but only in cosmology! By inventing a new "mysterious" particle or phenomenon, any needed characteristic can be ascribed to it. Only in cosmology.

Chapter 14

Arguments against the Big Bang

Problems for the Big Bang:

"Not a single important prediction of the theory has been confirmed, and substantial evidence has accumulated against it."

"Observations have been piling up for over 25 years and are now overwhelming."

Comments such as the above, and others in the "Selected Quotes Appendix," attest to the feeling of discomfort that many scientists are beginning to experience with the Big Bang scenario. In addition, more and more articles are being published in popular periodicals that reflect the sorry situation that cosmologists find themselves in. For example, here are a few *titles:*

Bang! A big Theory May Be Shot. [47]
Big Bang: Big Bust? [178]
Big Bang under Fire [50]

COBE Sows Cosmological Confusion [140]
Cosmology Theory Compromised [52]
New Challenges to the Big Bang? [87]
If not with a Big Bang, then what? [48]
The BB censorship [6]
The Big Bang never happened [69]
The Creation of BB bickering [13]
The universe may be younger than thought [66]
Universe Theory May Need Revision [193]

Weaknesses:

The realization of the Big Bang concept's weakness seems to be weighing heavier with each new observation, but the frustration of having to leave the security of a comfortable dogma, and the perceived loss of personal prestige for the most widely read cosmologists, make an admission of error painful indeed. But the drum beat goes on, and the evidence just won't go away.

Most obvious:

Galaxy Ages:

The extraordinary complexity and sophistication of galaxies are thought to demand eons of time, 15 to 18 billion years usually, for them to form, seemingly regardless of how they are supposed to have done it. However, the most recent farthest (oldest) galaxy is estimated to have formed within a few hundred million years of the Big Bang, and even that one is partially emerged in the dust of stars that had lived and died at even earlier times. No current scenario – no popular theory – allows galaxies to form in so short a time in a gravity mechanism. The problems in explaining galaxy development are so fundamental, cosmologists have had to resort to all of those exotic, strange, far out concepts that we have delineated ...again ...and again.

Galaxy Formation:

We have all seen pictures of bunches of galaxies all facing in different directions, but what ever wide-ranging force that could have resulted in the precipitation of galaxies would have produced neighboring galaxies all with similar centrifugal forces, similar angular momentum, and similar orientation! There is no way that nearby galaxies with different orientation could have been formed from the same particular event. The standard, well established, widely endorsed version of galaxies precipitating from shrinking gas clouds simply cannot work, ..as we've shown.

Globular Clusters:

The argument that globulars were thrown out of a galaxy because they were heavier can't be correct, because any mechanism for throwing anything out of a central body would have to be planar, much as we observe in our solar system, and the distribution of stars in the system would be largely based on weight.

Galaxy Movement:

Over and again, we are given examples of galaxies, stars, neutron stars even black holes that seem to have collided and formed some super-event of unusual magnitude. But how could bodies that are racing away from each other in an expanding universe ever collide? Next time that you get a chance to watch crews dynamiting earth or stone, see if any of the air-borne rocks or bits of debris can hit each other. Impossible! But, examples of collision are a minor consideration compared to the massive streaming movements of entire sections of the universe, flowing as if in rivers, every which way toward imagined great and greater attractors. Every Big Bang model simply ignores this contradictory evidence.

Quasars Ages:

Also estimated to be almost as old as the universe, which means they would have to have formed very quickly, but they are deemed to be much more complicated than galaxies and should take a lot longer to form. Their particular type of redshift is admittedly unreliable as distance, age, or energy determinants, and they might be quite near.

Structures:

The vast distances that galaxies and matter would have to travel to form these super-structures would require perhaps a couple of hundred billion years. And, no one seems to have any idea whatsoever why these galaxies and other bits of matter should even bother; yet neither galaxies nor atoms tend to form in sheets or crystals unless some kind of binding force demands such get-together. Explosive scenarios do not lend themselves to orderly arrangements of anything. Structures aren't needed in a Big Bang scenario, but the evidence for them, the redshift, cannot support the Big Bang if it can't support structures.

Small-Scale Inhomogeneities:

Scientist cannot explain how matter could clump into galaxies and structures of galaxies unless it is assumed that the beginning **was not** uniform. The question how these inhomogeneities could survive in a hot, gaseous, infernal, unimaginably dense and featureless, explosive soup is ignored.

Large-Scale Homogeneity:

Scientists cannot explain why the universe looks essentially the same regardless of direction. Considering that the universe developed so fast, densities should have varied and the further (older) galaxies and quasars should look different that the nearer (perhaps younger); unless it is assumed that the beginning **was** perfectly uniform. The question how homogenates could survive the trauma is also ignored.

Uniform Density:

A consistent density is observed everywhere, but volume expansion from a central point would dictate that the density of each farther, outer spherical layer should be proportionally less, e*specially* if it is assumed that the beginning was perfectly uniform. The raison pudding recipe for a universally-wide, universally-constant expansion ignores the evidence that there is no such expansion within individual galaxies or even within clusters of galaxies; and it ignores the realization that universally-wide constant input of energy is required to keep such a pudding expanding. See what happens when you turn off the gas under your pudding. The general uniformity of elements spread everywhere throughout the universe suggests an ongoing mix rather than a single event, as any housewife could tell us.

Gravity was late in clumping Matter:

Gravity, the very weakest of all the universal forces, in spite of the fact that gravitational attraction decreases by the square of the distance between masses, waited until density was at its least; until the masses were at their greatest separation, before gathering mass. The reason why cosmologists claim that gravity "waited," might be because they realize that gravity could not overcome anything but the weakest of [or perhaps any] gas pressures.

Inferred Expansion:

The *only* reason for scientist's insistence that there was a Big Bang is the evidence that the universe seems to be expanding and the rationale that it must be expanding from a point. Even if it is expanding, though, and it may well be in part, their favorite recipe for expansion is that of the pudding, which on our stove expands only when heat is applied and certainly not from any one point. Moreover, the only evidence that expansion is taking place is the redshift of light sources, but there are many arguments that redshifting is produced in a number of different ways, and that it is not a reliable indicator of expansion. This one bit of unreliable evidence is the sole reason that hundreds of bits of contradictory evidence are ignored.

Exotic Inventions Are Required:

Conjecturing and theorizing about possible particles and concepts is exactly what scientists should be doing as they try to deepen their understanding, but in cosmology, once conjectured, these possibilities become accepted as entities and support for other theorizing. The conjecturers and theorists have been mutually supporting each other to the point that the weaknesses of either are offered as a strengthening of both. Ergo Quantum Cosmology. The long list of inventions just discussed, are for the most part treated as fact, thereby inhibiting the thought-process and leading cosmology down the wrong road.

Singularities:

They are not relied upon by any scientists other than cosmologists, yet they are deemed crucial to the Big Bang theory. First, the big event itself, then the wriggling of strings which are merely another type of singularity.

Black Holes:

As discussed in Chapter 12, there *is no* conceivable way to make them, and evidence for them is terribly weak. Those few outpourings of energy that have been credited as evidence for black holes, were deemed so merely because cosmologists had no other explanation for the observed in a scenario that recognizes only the force of gravity. Black holes needn't, and probably don't, exist in any other model.

That Obscure Dark Matter:

Perhaps the saddest moment for physics was that when dark matter was invented. Merely because of the realization that visible gravity sources, by themselves, could not support the Big Bang theory, a much more powerful but unreal gravity source was imagined. The greater sadness is that most of the finest scientists seem to have completely accepted this bit of mythology rather than reconsidering their latest "absolute" understanding of the parts that gravity and electromagnetism can play in the universe. Unrecognized by any scientists other than cosmologists.

First Law of Thermodynamics:

The dictum that energy can neither be created nor destroyed has withstood every test, for all of time, except for that one Big Bang micro-movement. The creation of mass from energy fields in every other scenario, however, is very much in keeping with that law.

The Temperature of the Microwave Background Radiation:

From a temperature of as much as 100 billion times that of stars, which was to have existed anywhere from eight to twenty billion years ago, scientists applied their mathematics and observations to "predict" today's current background temperature of 2.73 degrees Kelvin. Such an offering is laughable. The now measured 2.73 degrees above absolute zero is every bit what one might expect for the deep vastness of an empty space warmed only by distant starlight, and what we can very well expect to last forever. As early as 1926, a temperature of about 7 degrees Kelvin for outer space "produced by the radiation of starlight" was proposed, but after the Big Bang proposal, "starlight radiation" was to be now called microwave background radiation.

That Damned Cosmological Constant:

Einstein's in and out "adjustment" has been tormenting cosmologists ever since it was first considered. It has been called: a *repulsive force* independent of matter to counteract *gravity,* a period *of time*, and popularly, *the rate* that the universe is expanding (but no one can agree on its value.) Supposedly, if 80% of the energy density of the universe can be attributed to the cosmological constant, it *might* explain the observations, but scientist can't even agree whether the energy should be positive or negative. ...But what the heck is it? It's not gravity, ..it's not electromagnetic. It is a *"conceived force"* not considered by *any* scientists other than cosmologists!

Omega:

Defined as the ratio of the actual mass density to the critical density needed to create a flat, delicately balanced between-open-and-closed, universe. With Einstein's value of zero for the cosmological constant, omega equals one, corresponding to a flat universe. But scientists tell us if that is so, space would have no energy, and we know that there is plenty. We can *see* some of it.

Disregarded:

Disregarded is the incredulously large discrepancy between the "measured" age of the universe and the very much older ages of galaxies, and the very much older ages of the so-called "structures" of galaxies.

Disregarded is the obvious unreliability of the redshift measurement.

Disregarded was the idea that the expansion, if real, could be explained by the normal fluctuations of a normal universe.

Disregarded was the fact that the finally detected temperature variations in that otherwise smooth microwave background do not meet the requirements for the much larger variations that large structures should demand, or that such temperature variations can be expected in even an infinity-old universe.

Disregarded is the observation that the universe is everywhere chock-full of teeming clouds of hydrogen atoms that "somehow" escaped the gravity that was supposed to have pulled almost all matter into the clumps from which galaxies were supposed to have been formed; or that those clumps must have formed much earlier than theories allow.

Disregarded is the realization that cosmologists have been utterly unable to assign meaningful values to their most crucial constants: The Cosmological Constant, The Hubble Constant, and Omega.

And, there is no end to that which has been disregarded, including the number of dissenters.

Suppression:

> There are a number of sad aspects to what we have been discussing here; certainly, one of them must be the discipline-wide collusion in keeping dissenting voices unheard. Cosmology is not alone in this sort of failing; but cosmology seems desperate in the effort. As Einstein and Maxwell and Kepler and others showed, sitting down and "thinking carefully" about a rather basic idea can open up new horizons, but rather than welcoming new ideas, cosmologists are accused of suppressing them:

Geoffrey Burbidge's scathing article in *Scientific American* accuses the community of censorship, even in school textbooks, and demanding that research proposals "toe the party line." "Conformity is the order of the day." He says that the Big Bang is intestable and is more of metaphysics than physics; that astronomers are restricted from telescope use if their findings contradict the Big Bang theory; that young astronomers dare not to offer innovative alternatives.

> "There is a heavy personal price to be paid by anyone that criticizes mainstream theories."
> "It is not so much that they are dishonest as that they take the path of least resistance."
> " Most scientists acquire incentives to keep a paradigm in place."[107]

> "Surely science would be better served if a greater attempt were made to loosen the grip of prevailing prejudices."[52]

> "They like the 'status quo.' There will be a lot of egg on the face. Peer pressure demands conformity."

> "It's not just a case of fooling the public (or themselves) it is case of inhibiting real science."

> "We hear of 5 or 6 classes of objects that violate the redshift assumption, but observations of these objects are banned from the telescope."

> "Most scientists are more interested in protecting the status quo, the current dogma, then they are in trying to explore new ideas."

As Captain Kkirk demonstrated, an ill-founded reliance on imaginative and exotic perception, and a formalized policy of restricting minority views, can negate the purpose of "the mission." Have our cosmologists given us the best results of their examination of the universe? If so, why the suppression of minority views? Einstein wondered if Farady would have discovered the law of electromagnetic induction if he had gone to college, but "since he wasn't encumbered by traditional thinking," he started to think in terms of "fields" as independent elements of reality. Has the profession been acting in accordance with the scientific method? If so, why is it so easy to demonstrate that they are on the wrong track? As I said earlier, my effort here is to convince you that there is something terribly wrong with Cosmology today, and offer the hope that we, at the grass roots, can nudge that profession back into the scientific track.

To be fair, we should recognize the human element in this. Why would scientists tend to favor the Big Bang scenario even when the evidence that it is a bankrupt theory is piling up? ...Well, it's naturally more palatable to defend an event, than it is that the event never happened. "Nothing happened" is not much of a headline.

Chapter 15

Fundamental Theorems and Concepts

---

* Measurements Are Inferred * Mathematics is Irrelevant * Logic Can't Be Ignored * Quantum Mechanics Doesn't Apply * Strings Are Immaterial * Relativity is Relative * Empty Space Isn't * Is Gravity Understood? *

Picture yourself as a graduate student in a physics class: where wormholes are described as popping in and out of existence, forming baby universes, and connecting us to parallel universes of up to 26 dimensions; a class in which false vacuums, antigravity, a cosmological constant that is anybody's guess, in a universe that is anywhere from 8 billion years to infinitely old, is tomorrow's required reading; and a class in which strings of infinite length and mass, uncertainty principles that tell us that we don't know where anything is, vacuum systems that can assume just about any energy – and do – are on today's pop quiz.

This is what physics has come to. It's not the physics that you and I learned when we went to school. Remember when we used to study acceleration, acoustics, Coulomb's Law, cyclotrons, elasticity, electromagnetism, …energy, and a host of other "real" subjects? These were down-to-Earth subjects that still do matter on Earth, and even in our solar system, but not, apparently, in outer space. What we had was "classical" physics; not less sophisticated, but of time-honored procedural methodology. A system of theorems, testing and proofs that left no doubt in anyone's mind of their validity. …Not so now.

On Measurements:

Cosmologists (because they deal in the farther out) have far more difficulty than Astronomers in making measurements because the distances of interest are usually well beyond what parallax (trigonometry) allows. Surveyors use trigonometry all the time, and as you know, the farther apart (the longer the base) angular measurements are made on the same distant point, the more accurate the determination. The longest base available to cosmologists that can be attained for making angular measurements, is the distance from one side of Earth's orbit around the Sun to the other, and about the farthest distance that scientists can measure with this longest available triangulation base is around 2000 parsecs, or 6000 light-years, 6000 light-years, though, is only of limited help for determining distances in a universe said to be in the many billions of light-years wide.

You recall how astronomers have cleverly stair-stepped their way out to the fringes of the universe by basing their farther estimates on their closer-in estimates. Their ingenious detective work of using stair-stepping "standard candles" is perhaps the most often used procedure, other than redshift measurements, for estimating distances beyond the 6000 light-years. Starting with an estimation that certain star types have similar intrinsic intensities, and judging their distance by their observed apparent intensity, distance is inferred. When members of that first type are thought to be near members of a different, farther, brighter type, distances to members of the second, farther type can be estimated. And so on.

RR Lyraes, Cepheids, and Ia supernovae are the so-called standard candles. The average brightness of RR Lyrae stars was calculated based on light colors of stars similar to our Sun. The Cepheid's average brightness was inferred from that of the RR Lyrae's oscillations estimated to be related to intrinsic brightness. The Ia supernova's average brightness was based on the Cepheid's. While parallax measurements are reliable only to about 2000 parsecs, we use RR Lyrae stars out to about 200,000 parsecs and Cepheids out to about 4 million. Clever as all this might be, there are many estimates based on estimates based on estimates involved.

As it turns out, RR Lyrae stars oscillate in size and vary in brightness. Cepheids come in two types and of different intrinsic brightness which can vary over a very short period; and they can vary in luminosity depending on whether they are found in galactic centers, spiral arms, or out in globular clusters. Supernova Ia's are believed to be at the same brightness, (not all scientists agree) but no one knows what that brightness is. And, dust, light diffraction, and perhaps even undetected lensing all add to the unreliability of estimating *any* brightness, or distance. But, ...what's a cosmologist to do? There are a few examples of measurements that are reliably accurate, such as the expanding Crab Nebula and Supernova 1987A, because we know when they happened and can reasonably estimate their expansion, but they really don't help our stair-stepping estimation efforts.

> "Thus the final estimates...are contingent on the correctness of a long chain of assumptions, each one contributing significantly to the possible error in the final result."[238]

Hubble's estimates of distances based on redshifting were too small, because the earlier period-luminosity relations of Cepheids were later determined to be incorrect. The Hubble constant, which is used to determine recession rates, has been decreased by scientists a number of times over the years, which tends to make the universe seem older. We have discussed age determinations well enough already, but it must be realized that even the hallowed sources might be revised or later considered incorrect.

That crucial stair-step benchmark, the Virgo cluster, turns out to be not as old as thought; it is now measured to be even younger than the galaxies in it! And, of course, it emphasizes the tenuous nature of any of these measurements. On top of this bit of quandary is the realization that the accuracy of even that recent measurement can be questioned, because there is no way that scientists can tell whether the marker stars used in the measurement were in the front, center, or back of the cluster!

> "It must be emphasized that all of these methods of investigating the structure of the universe have built-in assumptions, which may very well be incorrect."[237]

Now. Having said all that, let me back-track a bit. All of the stair-stepping estimates that astronomers and cosmologists have been making all of these years to infer the deeper distances has indeed been a clever effort. And a lot of the perceived relationships of the near to the far are probably valid, but we must not forget that there are *no* absolute measurements of anything beyond the fringes of the Milky Way. Cosmologists like to apply quantum mechanics to the universe proper, but perhaps *only* the renowned, absolute, now and forever *uncertainties* of the quantum world are appropriate for cosmological considerations.

"So, at the present time we must limit ourselves to computing probabilities. We say 'at the present time,' but we suspect very strongly that it is something that will be with us forever – that it is impossible to beat that puzzle – that this is the way nature really *is*."[57]

When we have run the standard candle stair-stepping out as far as it will go, the only thing left is redshifting. The ratio of redshifting to distance has been popularly agreed on, but we have shown, I think, that it's reliability is not at all that certain. Everything from quanticized measurements, to bridges between bodies with very different redshifts, to heavily redshifted nearby bodies, add up to the realization that not even redshifting can be relied on for determining distance. At the present, *all* estimates of distances beyond 2000 parsecs, or 6,000 light-years, is iffy, and considering that the Milky Way has a diameter of about 10,000 parsecs, or 30,000 light-years, we shouldn't put much stock in universal measurements.

There is no meaningful evidence that redshift is related to velocity;[106] and it could be caused by gravity,[74] or characteristics of space that we haven't noticed yet. One author lists 20 "non-velocity" redshift mechanisms, and there are many comments noted in the publications that reflect on the unreliability of the redshift device:

"Almost every super-giant in both the small Magellanic cloud and the large Magellanic cloud have 'excess' redshifts, and stars right here in our own galaxy have shown excess redshift."

"The spectral pattern of stars seems to change with the Sun spots."

"Even partially coherent sources can display changes in redshift and might be useful in communications."

"The Royal Observatory verified Tifft's ten-year-old periodic of galactic redshifts."

"There are examples of bodies of very different redshifts being connected and therefore at the same distance."

"The evidence is piling up. More examples of redshift discordant pairs of galaxies keep surfacing."

It might be quantesized; as much as 15 or more miles per second."

"It might change with time."

"May depend on the type of galaxy."

"Spirals tend to have higher redshifts than ellipticals in the same cluster."

"There is a good correlation of redshift and apparent brightness for galaxies."

And, ...William Q. Sumner is quoted in a recent *Meta Bulletin* that it was shown in 1939 and confirmed in 1958 and 1994, that an electron's orbital parameters change in an "evolving" universe: that atoms change with geometry. A significant implication of this finding is that redshifting might mean not that the universe was expanding; but that *it was contracting*! [126]

Redshift is the crucial determinant in the discovery of the great walls of galaxies that shed so much doubt on the Big Bang. But! It turns out that relative movement of galaxies many have caused erroneous estimates of structure arrangement; that perhaps they aren't so great after all. Perhaps those great walls don't exist! Would this consideration favor the Big Bang? No, because if redshifts are unreliable in this instance, they can't be counted on in any.

If redshift is unreliable, so is:

Distance.
Brightness of the source.
Energy of the source.
Expansion.
The entire Big Bang concept.

AND, ...even if redshift *does* work, even if the universe *is* expanding, it still does not require a Big Bang!

On Mathematics:

Almost every single thing "known" about the universe beyond our local galactic group is derived, or contrived, from mathematics. Cosmologists, most of whom seem to be Big Bang enthusiasts, rely on mathematical constructions rather than laboratory experience, partly because they have no choice – direct observation and testing is impossible – and partly because they have tied their wagons to a falling star (the Big Bang theory) and have no choice but to try, at all costs, to prop it up. Mathematics has been called "the last resort," recognizing that when physics can't defend a concept, mathematics is asked to. It has also been said, "mathematics belittles intuition or experience."

It has been estimated that about 10% of astronomers sit around developing simulations of the universe on computers,[42] and results are periodically announced as new "models" of the universe. Invariably, though, aspects of the models are soon shown not to agree with observation, so most of the models go through a series of modifications that attempt to correct the discrepancies. Much too often, when the models are found to be insupportable on their own, mathematical "discoveries" of new features of exotica are applied: new kinds of dark matter, black holes that operate differently than run-of-the-mill black holes, new uses for cosmic strings, new "different" particles, and ...etcetera. Too frequently, scientists, humans that they are, are caught up in the search for evidence that their favorite hypotheses is correct, and if exotica will make the model work, ...well, what are you going to do?

And, favorite hypotheses (bias, really) are *required* before a computer program can be designed! A scientist cannot have an open mind when he designs a program! A preconceived notion of facts is necessary before one can lay down the parameters of search! You have to know what particular data you are looking for before you can tell the computer what to look for. For example, if you want to detect 10-meter wavelengths, you design a 10-meter antenna. And it won't pick up anything else. Antennas, instruments, computer programs; all are designed and selected for very particular purposes. You want to detect neutrons? Stick a big tank of cleaning fluid about a mile underground, but don't expect it to detect gravity waves – or anything but neutrinos.

Most of the time, the problem is merely one of detail and data. To develop an accurate program that can describe universal events, far more data and detail than can possible be gathered, and a computer with far more capability and speed than has been demonstrated so far, are required. Shortcuts are essential. For example, it has been philosophized that a butterfly's flapping in Peru can initiate a tornado in Topeka. Maybe so, but design a program to prove that one! And that's just the atmosphere; the universe is supposed to be more complicated. Another problem, is that each new, bigger and better computer and program that is used produces results quite different from earlier efforts. Brave souls can only use estimations and simplifications for their inputs and hope for the best; then a series of adjustments and corrections have to be applied to make the thing come out right.

"It is evident that any single experimental fact can be disregarded if we are willing to think up a special reason why the experiment should show the result that it does."[228]

It is a given, that without an idea, a structure or theory of what to look for, you simply won't see it. If you are looking for locomotives, you won't see the deer in the woods. Pre-conceived notions, and the dogmas of religion kept astronomers from seeing anything but Aristotle's concentric crystal spheres for 2000 years. And if your idea, or evidence, or observation conflicts with the popular view, don't expect any funding. Just as astronomers should be embarrassed by the metaphysics of cosmologists, mathematicians should be embarrassed by the credence that cosmologists give to recreational numerology. Computer models, and mathematics, are not always the answers.

No set of laws has governed man more broadly, more pervasively, longer or better, than the laws of mathematics. Mathematics has ruled (and freed) man more inclusively and completely than any other system of human discourse; so much so, that to say it is to be redundant. From the first tallying of grain and fish, to the precise trajectories of cannon shell, to the exact determination of satellite movements, to the exquisite forecast of planet positions and solar eclipses; mathematics has become a part of us, and perhaps even a part of the heavens themselves. All of the motions of the Sun and the planets have been measured, tested, and authenticated by thousands or astronomers, for thousands of years, to the point that we are all in complete agreement as to where Mercury and Pluto will be a thousand years from now. Who could doubt that mathematics is ubiquitous – even to the far reaches of the universe – from the beginning to the end?

The mathematical determination of the energy gained or lost by an electron as it moves from one discrete atomic level to another is undoubtedly the same on this side of the universe as it is on any other side, for all time. And the same can be said for the actions of the atoms as they enter or leave molecules. But what of the molecules, and the particles and bodies and stars and galaxies that they constitute? Mustn't mathematics divine their comings and goings throughout the universe as well?

Maybe not!

Mathematics, especially now that we have relativity, can precisely predict the present and future motions of the planets, because their past motions have been determined with an outstanding accuracy, and gravity's influence on them is pretty well understood. Gravitational forces within the solar system, that is. Mathematics can predict motions only when extensive experience with, and complete understanding of, the forces involved are available to us, but this is not so for the universe at large! Even the cosmologists admit that they don't know for certain whether there really is gravitational dark matter floating around out there, which kind, or how much. Distances are educated guesses; whether the universe really is expanding or not is still unknown. We don't know how the weakest force in the universe – by 40 orders of magnitude – works in conjunction with electromagnetism; and, there is credible reason to believe that gravity has a limited range after all.

> "Any vague theory that is not completely absurd can be patched up by more vague talk at every point that brings up inconsistencies – and if we begin to believe in the talk, rather than in the evidence, we will be in a sorry state."[232]

Well then, if the same time-tested certainty that has been developed for planetary motions within our solar system doesn't exist for galactic motions in outer space, then there is no way to apply mathematics to their predicted movements. Of course algebra, calculus and your calculator will work beyond Virgo, but all the mathematics in the world won't help if you don't know what it is that you are counting. ***Mathematics, then, doesn't count outside the solar system!*** And if this is so, then just about every cosmological idea that is mathematics dependent, and just about every one of them are, are meaningless! Schoolbooks and television programs present information about the universe as facts, even though only opinion and conjecture are available to the cosmologist.

> "We must be careful to interpret the results of our theories when they are treated with full mathematical rigor. We do not have the physical rigor sufficiently well defined. If there is something very slightly wrong in our definition of the theories, then the full mathematical rigor may covert these errors into ridiculous conclusions."[231]

One of the very latest branches of mathematics to be applied to physics is that of Chaos, a system of equations that predicts the future from past events. Convection in fluids, laser and electromagnetic oscillations, celestial mechanics; even weather forecasting now use the very handy system of chaotic dynamics. Chaos deals with system patterns of behavior called attractors, but more than one usually renders the results unpredictable; the slightest change, or unforeseen or unobserved happening, and you can throw the prediction out the window. If that Peruvian butterfly happens to be a bit lazy, that Topeka tornado might turn into a Bangor blizzard. But it is one thing to try to forecast the future with mathematical descriptions of measured and analyzed processes, and quite another to try to determine initial situations from observed results, knowing all the while that *any* perturbation can render the "forecast" meaningless. It follows then that it is wishful thinking at best that current events or situations can be traced back to a starting point in a system as complicated as the universe. Today's redshifts do not a 20 billion-year old Big Bang make.

The Big Bang is almost entirely a mathematical construct. And this is not your average mathematics! Scientists, in an embarrassing display of desperation, latched on to that extraordinarily complex mathematical contrivance, the Higgs Field, which is supposed to allow inflation to work! Supposedly, "it would transform *any* sort of bizarre, irregular non-uniform Big Bang into a large and smoothly expanding universe."[135] Sheldon Glashow, a prime user of the Higgs mechanism, is credited with referring to it as "the toilet (the necessary room) in the house of physics." [131]

Mathematical extremism is required where we approach the extreme: the extremely small point from which the universe is said to have been born, from nothing; black holes into which everything is to fall, or an infinite number of unattainable, unknowable universes. What is missed in mathematical musing is the realization that larger-scale influences of nature and Order *don't let the processes get to those extremes.* If too many deer are eaten by wolves, the wolves die off from hunger and the deer come back. If a vacuum develops, gasses and matter rush in. If the pressure get too great, something gives. It's merely a question of keeping your perspective.

Mathematics is an exquisite, absolute "truth" unto itself: it is a tool. Absolute, though? Not really. Mathematics is really a language, and we all have to speak in the same tongue to understand each other: 10+2=12, but 10+3=1, and 10+101 =7, right? *Yes,* but only if we know which dialect, or in this case, which system, is in use. Mathematics is a truth? If so, so is Spanish or French. Well, it is certainly a tool, but it can be misused just like the sharpest saw can cut a board to the wrong length; or the brightest surgeon, operating without the slightest loss of blood or leaving even the tiniest of scars can cut off the wrong leg or operate on the wrong person. Mathematics may well be a "last resort." But it is not physics. More so, it certainly is not cosmology.

"Einstein wondered why physics should be inherently mathematical."[127]

"David Lindley noted that many sciences make use of mathematics, but only theoretical physics is thoroughly steeped in it."[128]

On Logic:

All theories and concepts have a shelf-life. Back when the Greeks first started thinking of these things, the universe was thought to be only a few thousand years old. Now the estimate is between 7 and 20 billion. It seems that as science matures, estimated universe ages get older. Why doesn't it follow, logically, that we'll be a lot more sophisticated when we estimate that the universe is infinitely old?

Scientists do not readily admit to the use of logic, because the term implies judgment, and true science is not supposed to be judgmental as much as it is experimental and causal. But logic is not the same as "common sense." Logic: the solving of syllogisms, is indeed mathematical; but scientists realize that the best logic can be faulty if one or more premises of the syllogism are faulty. Be that as it may, logic is not all that foreign to science: logic – even common sense – is applied to observation and test results in spite of best efforts to avoid it, because the rawest data must be put in perspective if it is to be communicated and applied. This, then, is why those 40 scholars, scientists, and other experts felt that they had to defend the scientific method, and urge a return to reason *and logic*.

The poor cosmologists have almost no choice but to try to apply logic, judgment, and unavoidably, common sense, to their observations; because they have little else. Almost all their data is questionable, and there are almost no tests that produce concrete results. It is inevitable then, that even though they share the same data, cosmologists suffer many different conclusions. Equally highly trained scientists frequently differ in their interpretation of observation, and the future offers precious little hope that we will ever find out which, if any, is correct. Logic might have it then, that if scientific experience is not a determinant for applying logic, or common sense, or reason to what we observe in the sky; you and I might as well try our hand at it.

* * *

It came to pass one day that the Mayor called in his Chief Fixer-Upper, and said, "I am going to be away for a couple of weeks, and I'd like you to look into the swamp on the north side of town to see if it can be developed."
"Will do, Your Honor," said the Chief Fixer-Upper, and saluted as the Mayor went off into the temporary sunset.

The CFU immediately went to Flow Control and presented the problem to the flowologists.

"The first thing we need to do," responded one, "is to do a mathematical analysis of the flow of water in the swamp that we might better determine what can be done."

That's easy," said the second flowologist, since there is no flow, $F = 0$."

"No flippancy, if you please," said the first, "because the flow of water is a lot more complicated subject than meets the eye. Water invariably flows down hill, or course, and water molecules falling from a height experience turbulence and agitation; kinetics, chaos, and a number of such applications may be required."

"That sounds like a better approach." Said the CFU, "in this day of environmental concern we've got to dot the I's and cross the T's. Let's treat this professionally."

And so, Flow Control did develop unto themselves an exquisite mathematical analysis that involved chaos; Fourier, Laplace, and Lorentz transforms; and the CFU was pleased, just as the first flowlologist expected he would be.

The CFU took the results back to his office and looked at them. "We have a true mystery here that needs further in-depth study."

And lo, the Mayor did return after his allotted interval. As he passed Flow Control, he saw through the open door that nothing was happening at all, but one flowologist, sitting there by himself, looked up and said resolutely, "$F = 0$!"

"Thank you," said the puzzled Mayor; and he walked on, only to pass an obviously very busy CFU's office; over the door to which was mounted a nicely printed sign: "Office of Missing Waterfalls."

The CFU beamed as he looked up from the work that he and a group were working on, and said proudly,

"We're hard at it, Your Honor."

* * *

Exquisite mathematics can show an impressive analysis of water molecules experiencing the chaos of tumult and turbulence in a waterfall birth, and that report might be much more interesting than one on a sluggish trickle, but it might not be appropriate.

It's easy, also, to envision a parallel happening in which a raft full of physicists are studying, in exquisite detail: the water flow, temperatures, salinity, clarity, and other crucial characteristics of a river ...as they approach a fall – and no one is looking out ahead. This little bit is not to make fun of scientists, but to point out again, that even the best scientific efforts can be swamped if complexity is valued more than simplicity. Even the best mathematics can be poorly timed, out of context, or misdirected.

Quiz time:

> Theory A requires exotic, weird, undetectable considerations.
> Theory B does not.
> Which theory is most likely to prove correct?

The exquisite and tedious analysis of the micro-events of an instantaneous primeval raison by eminent and ingenious scientists seems an exercise in futility. What good is it, other than to serve in mutual support of quantum and Big Bang theories that are understood by precious few? No new laws can be developed, certainly none that we can use, because the conditions will never be repeated, if they ever happened at all. And we can never know; because if it did happen, it was supposed to have happened in the dim dark past; and none of the findings can be tested.

On Quantum Mechanics:

A typical method of studying subatomic particles involves bouncing photons off them in huge "accelerators" and analyzing the results. The trouble is, the particle that we might be studying now is so small in size and mass that the photon itself disturbs that particle, and it is then not as it was. The Heisenberg Principle of Uncertainty recognizes this and states that we cannot accurately measure both the location and momentum of a particle. This simple statement is regarded as one of the most profound of modern science, and it is the foundation of a thought-process that influences all of quantum physics and the more pertinent aspects of cosmology; cosmology, because it has adopted a great deal of quantum theory to support itself. It's not the only principle of quantum theory by a long shot, but it gets an awful lot of attention.

A serious trouble for science might be that the constraints of that principle have been extended to the point that scientists seem certain that they can't determine very much at all in quantum physics or cosmology, with certainty. For example, even though the search for reality is fundamental to physics, the Uncertainty Principle has led quantum theory to deny any absolute reality at all! They tell us that there is *no such thing as basic reality!* Any system can assume any energy, including vacuum systems. It doesn't allow for infinite subdivision. Space seems to have 26 dimensions! Not 27, no, no: 26. And, of all things, the singularity of the Big Bang couldn't be, because we couldn't be certain of its location! (Ignored by advocates.) Quantum physics, then, solves the problems by making everything fuzzy: or as someone said: it is "A sophisticated, complicated, incomplete justification of an aesthetically pleasing preference." Mathematically, quantum physics can justify just about anything desired: infinite speeds, warped space, wormholes, seeing into the future, time travel; you name it.

Uncertainty tells us that there is no way that we can tell when a particular atom is about to decay. This seems understandable, and unequivocally true today, but, couldn't this be more a reflection of our present inability to measure such things now than a forever unfathomable mystery of nature? Isn't it possible that the profound Uncertainty Principle is really a manifestation of our present limitations of capabilities than an eternal, universal dictum? Today's inherent uncertainties of nature, such as the unpredictability of an acorn's fall or a dandelion seed's flight, might indeed be made more certain when oaks and dandelions are grown in a more controlled horticultural environment.

Along with the Uncertainty Principle, it has been determined that there are three, and only three, families of fundamental particles that include quarks and leptons, electrons and taus, and others. Most had been detected a while ago, and just recently the last "confirmed" particle was the "top quark" which is said to be bigger than an atom of gold! This is all well and good, and applause is deserved for quantum physicists for having reached this level; but ...is this really the ultimate? Isn't this *a* level, instead of *the* level? Even though photon bouncing (and the Heisenberg Principle) doesn't seem to permit measuring to lower, smaller, even more basic levels, why should we assume that future procedures wouldn't detect deeper discoveries?

The top quark, for instance: bigger than a gold atom. Bigger than a gold atom! How could any one, mathematics or no, accept a particle of that size as fundamental? What is *it* made of? Is it layered? Does the top half differ from the bottom? Does it have a surface texture? Mustn't it be made of still smaller particles? Questions like these must be asked of every particle, quark or no, over and again. Mustn't that very large quark be made of still smaller particles? Of course. And, ...wouldn't you know, yes indeed, a sub-quark particle has been detected, and ...it's back to the old drawing board. Even photons, thought of as bosons, or force-carriers; supposedly fundamental bits of energy, can now be split or given different polarities. Just to confuse the issue, a photon can convert to an electron and positron! But these are fermions, (and anti-fermions?) which are particles! Whether even the photon has a sub-harmonic, a smaller, more basic element or not, we may never know, but the very smallest particle of matter (if there is one) must be, rather a particle of energy!

Just as you and I can envision each and every particle of matter being composed of ever-smaller particles of matter, down to the infinitesimally small; so can we intuitively envision even-so-called fundamental energy packets as being divisible. Isn't it possible, and acceptable, that these energy packets are something like a resonant frequency? A "tuned" tone or cord of nature that has a sub-harmonic or even more basic forms?

The electron, for example: it sometimes acts like a particle, and it sometimes acts like a bundle of energy. When it is running around in space outside the atom, it seems every bit the particle, but when it is firmly settled into an atom, it acts more like a resonant frequency. While it is a particle, is it sub-divisible? Shouldn't it be? The point here is that uncertainty principles, or exclusion principles, or whatever, profound or not, might merely be the mode of the moment, and they need not be ultimate constraints of anything! All through the history of science we have had noteworthy pronouncements of finality that simply didn't last. What this amounts to, I guess, is that good old classical physics should not be dismissed at either the quantum or universal levels, and the exotic aspects of particle theory are suspiciously rickety support for an invented Big Bang scenario.

There is no question, whatsoever, that particle physicists have made great, concrete strides, but the future for their efforts is being lamented. To go further, far greater, far more expensive accelerators would be required, and the money just isn't there. The effort to find the top quark took 440 physicists six years to build the detector, which weighs 5 tons and stands three stories tall.[151] (Congress, you remember, killed the Super Conducting Super Collider.) Life is getting more complicated; every new effort seems to come up with a new particle that has charm or spin or strangeness, or other deviations that make little sense. Over 60 different sub-particles do seem to have been found in accelerators, a lot of which were truly predicted, but there are many predicted, or hoped for, particles that have not been found, and perhaps never will be. History tells us that if these "needed" particles aren't found scientists will simply tinker with the theories rather than discard them; they never discard a theory, anymore that we can discard government programs.

But most physicists don't understand quantum physics! Quantum physics, like all specialties, specializes. Only the special few in any specialty are up to snuff in that particular special consideration. The rest of us merely accept their findings as unequivocal. NO ONE has the time or inclination to question the fundamental findings of ANY profession. Including most of those in that profession! You and I don't question what the doctor tells us, and most doctors don't question the medical researchers! So the findings of quantum physics are accepted by the cosmologists, and vice versa, such that the uncertainties of one group are used to reinforce the unknowns of the other. That quantum theory can be applicable outside the atom and in the far reaches of space is *by agreement*! Since both quantum physicists and cosmologists have reached their limits of experimentation, they *agree* to use each other's concepts as *solutions*!

> "...I would like to suggest that it is possible that quantum mechanics fails at large distances for large objects."[229]

Two of the great theoretical physicists, Stephen Hawking and Roger Penrose, have been debating: the niceties and fine-tuning of the conceptualized experiences of "information" in black holes, the symmetry of quantum physics in reverse time, the natural curvature of space in the absence of matter; and other aspects of quantum theory beyond, perhaps, the off the street scientist.[102] Which of the two is more correct, is not the issue – for a couple of reasons. Most of us can only judge the quality of the arguments, but not their accuracy; and it is not at all certain that black holes even exist. More significant; the fact that two of the greatest minds interpret the same mathematical constructs differently, reminds us that the exactness of mathematics has to be interpreted philosophically! It is reason and experience that prevails, not the language; and that's what mathematics is: a language.

On Cosmic Strings:

We have made the point, I think, that even though singularities might be fascinating for mathematicians, they are not very meaningful in real-life, because they can take on any value and can't be applied to anything is nature. To circumvent this verity, cosmologists developed "stretched" singularities – singularities with no dimension other than length – and they were off and running. It stands to reason, that "string theory" can be very handy in solving mathematical problems, and there can indeed be applications in real-life, such as in the study of liquid crystals or plasma; but there too, we want to remember that while mathematics might be helpful in understanding certain phenomena, and that mathematics may indeed mimic nature, nature need not mimic mathematics.

But string theory was like manna from heaven for cosmologists that desperately pick up any crutch that might get them by the disruptive roadblocks that periodically, inevitably, tend to derail the Big Bang train of thought. Super-string theory postulates that the basic building blocks of matter are not point particles, but one-dimensional "strings." Beautiful. String theory seemed not to be susceptible to the misgivings that singularities generate; instead, scientists in the large have eagerly, and even productively, been applying the methodology in many different applications. Cosmologists, however, have not been treating string theory as a device for studying the universe, rather, they have adopted the concept *as an actuality*: happenings don't just perform in a manner described by strings; things happen *because* of strings! Indeed, they are supposed to be remnants of that false vacuum that we talked about earlier; they are given credit for the seeding and sowing of dimensional structures of galaxies from out of empty space; they are supposed to generate gravity waves, and they are the acknowledged force that launched the Loitering model universe from its lethargy.

Depending on the equation used, strings can be just about anything: one-dimensional, extremely thin faults of spacetime that are infinitely long, threading the entire universe from on end to the other; they might have no ends but are closed loops the size of galaxies; or they can be infinitesimally small, some 1200 billion times smaller than a proton, and make up the entire universe. Particles are said to be nothing more than balls of strings. They have almost infinite energy, vibrate fiercely, and have a thousand trillion tons of mass for every inch of length. "Strings can turn into black holes and black holes can turn into strings."[201] Tension is said to make them whip back and forth at near the speed of light, frequently crossing one another, breaking up into ever smaller loops, generating gravity waves as they shrink and disappear. (But, they may have survived in their original state instead of cooling off like the rest of the universe!) And, ...information is supposed to flow along a string at the speed of light.

String mathematics does seem to be fascinating, and perhaps useful, but is almost irresponsible to claim this discipline, or any other, as proof that cosmologists have the right handle on the universe. Strings, or quantum physics, might indeed help the thought-process, but they are, after all, some of the most sophisticated, state of the art mathematical exercises, and only the most dedicated, capable mathematicians can fully grasp them. The theorists that work on these equations are so involved in the difficulties and niceties of the mathematics that they are not really in a position to apply them to the world of practicality, and usually don't try to. But cosmologists do.

These devices are invariably taken as gospel, as entities, in spite of the fact that most physicists cannot hope to understand their intricacies; they have to take them as is. And they do, with relish, in spite of the fact that there is no evidence whatsoever that they are accurate descriptions of the universe, or that they most certainly are not *causes* of observed results.

Dark matter theories, as you remember, hold that the universe is homogeneous and variations are the same everywhere; but strings whipping violently through space, breaking up into gravity wave generators, should surely have produced a more chaotic situation; the enormous mass attributed to strings would distort space a heck of a lot more than anything that we have observed, and if violently wiggling strings would precipitate galaxies, the universe should be solid with them.

> Taking a breather, for a moment, if strings can precipitate galaxies and walls of galaxies, why do we need the collapsing-cloud scenario of making them?

And, by the way, if the energy of strings goes into the making of galaxies, what happens to partially depleted strings? Do they slow down their wiggling? If they can't make galaxies anymore, do they stop expending energy – and survive? If so, why are none detected? It's a cinch that the Cosmic Background Explorer satellite (COBE) has detected nothing in the microwave background radiation that could be attributed to strings. Another little conundrum for string fanciers is that the theories are consistent only if strings originally inhabit a ten dimensional spacetime, but the theories are marred by theoretical particles theoretically exceeding the speed of light. Wait now, a universe of 26 dimensions seems to cure *that* problem, but nothing in string mathematics indicates a preference for our familiar four dimensional universe, or even considers it any more likely than a two, thirteen, or 20 dimension universe.

In pursuit of meaning for mathematical string theory as pertains to the universe, some scientists have made a thorough study of the inhomogeneous stress reactions of liquid crystals and friction-free super-fluids. By applying and removing a steadily increasing pressure, line-like discontinuities that change with pressure appear in the otherwise clear substances. They equate them to the so-called symmetry-breaking phase transitions, and perhaps even monopoles, that are said to characterize cosmic strings and how they are supposed to function in the universe. This is another ingenious example of how scientists cleverly go about their trade, but here, shouldn't we have to ask ourselves if string theory mightn't be more applicable to liquid crystals and super-fluids than it is for the cosmos?

No one has yet come up with a test for string theory, and there is no hope held out for one, ...but it doesn't matter; ...it's needed for the scenario.

On Relativity:

Relativity. Relativity is perhaps the most elegant of the theories because of it's beauty, the "feel" of it, the way that it made new use of the mathematics; the way that it led the user to step back from established preconceived notions. It has tested well, in spite of the fact that it hasn't been tested thoroughly.[11] Its mathematics is full of varying variables, and what analysis has been done has usually relied on quite a bit of simplification and assumption. Regardless of the questions that still remain, relativity has established some absolutes that have been, for the most part, generally accepted: that nothing can exceed the speed of light, that space is curved by mass, that mass increases with velocity, that everything is relative; et at. Beautiful.

But Relativity is a gravity-field thing. It does indeed describe and define the effect that gravity has on measurements of distance, direction, or time in a gravitational field. But, where gravity approximates zero and temperatures are uniform, such as out in the deeper parts of empty space, good old Euclidean geometry should work just fine.[233] One scientist says that he can derive a complete gravitational dynamics from Newtonian interactions in flat spacetime.[252] And, one great man says, "In fact, after more than 50 years in the business, I am just beginning to realize curved spacetime and relativity is just pretentious, superficial and irrelevant.[275]

Even this most accepted proclamation of finality might have a shelf life. History tells us that this must be so. (The first thing that Big Bang enthusiasts did was to throw it out the window.) The Meta universe, by the way, explained some of the features of relativity in classical terms and showed that some things might indeed exceed the speed of light.

Friedman showed that general relativity didn't allow for a static universe, but that it could be EITHER EXPANDING or CONTRACTING. Well, now, even here in the consideration of this universally accepted theory, might not the precision of the mathematics be misinterpreted? Couldn't it be applied to a universe that is expanding in one area while contracting in another? Or one that expands at one phase and then the collapses in another?

On Empty Space:

We are not talking about the universe here, we are talking about the emptiness between things in the universe: "empty" space. Space is held out to be pretty mysterious stuff: quantum cosmology says that there is a probability that a small chunk of spacetime can just pop into existence ...from a wormhole. It can explode, be torn, expand, curve, and harbor vast amounts of energy; quantum theory not only holds that empty space is not empty, but that it is literally bursting with energy; so much so, that it amounts to 10^{80} ergs per cubic centimeter! The vacuum is said to be thick with virtual particles so much so that they should warp space independent of matter.[31] It has fabric; it's a sea of "randomly fluctuating electromagnetic fields." It is said to have density, texture, mass, substance; up to 26 dimensions, but no meaning if nothing is in it. Einstein said that the concept of space detached from any physical content does not exist. "It has active properties: it is not inert." All of the above pertains to that very emptiness that you and I look through when we view the moon, the horizon, or this page in this book.

The idea of space itself expanding takes a bit of getting used to. We are told that bodies are not merely separating, but that a cubic meter of empty space is growing larger! In deep space, that is, out beyond the galaxy; not inside the galaxy, mind you, just outside. That it expands is a necessary characteristic of a Big Bang theory. "Space keeps popping up between things that are separating." We are not just saying that the density is decreasing, but it is. Supposedly. Density is sometimes blamed as if it were a force. If density is the cause, then the flatness of space is indeterminable because we can't tell what a natural density is. A changing density is proposed in some scenarios; an idea more appropriate for an infinitely large universe of different areas at different densities, but density changes aren't what we are talking about here either.

The idea of an empty anything expanding makes little sense, but of course, space isn't supposed to be empty; it is supposed to be chock full of energy, texture, ...substance. But if space can explode and expand, what moves it? We understand how gas expands, but "empty" space? If space is expanding, what expanded it? If more space is being introduced to the universe to explain the expansion, it must be introduced everywhere, universally, simultaneously, pervasively, uniformly, homogeneously. Not in lumps; no, no. A ubiquitous, constant introduction of new space to old, everywhere. No mention is made of where this new space comes from, no theory explains the why,

or whether additional, "new" energy comes with it. Einstein did say
that the concept of space detached from any physical content does not
exit, but if so, was this "new" space associated with physical content
before it was introduced to the old? If it is being introduced everywhere
in the universe simultaneously, is the new expanding along with the
old, from the same Big Bang point? ...Does anyone really believe any
of this?

The concept of space itself expanding results from the logic that the
farthest galaxies are moving away the fastest and there is no reason that
they should. The raison pudding bit. An expansion of the universe, or
parts of it, does not offend the senses ...or classical physics. Classical
physics is quite satisfied with the idea that bodies move in relationship
to each other, such that the surrounding and in-between space seems
not to expand, or be effected, or even care. But the concept of "empty"
space expanding, an idea crucial to the Big bang theme, just cannot be.

On Spacetime:

Have you heard of the miraculous liquid that is the answer to a
housewife's prayer? It cleans clothes, refreshes your face, dissolves
stains, makes excellent soup, and allows plants to bloom. It also
conducts electricity, puts out fires, quenches your thirst, cures
hangovers, and flushes toilets. ...If water can be made to seem magical
and mystical, why can't spacetime?

Spacetime. What an exotic concept! A consideration that is made
out to be something magical, mystical, mysterious, ...significant;
something only cosmologists are privy to or capable of understanding;
the explanation for a lot of unanswerables when black holes don't help.
Spacetime is, though, nothing more than the three dimensions of space
that we're most familiar with, and the fourth dimension of time that
Einstein added for us. For you and I, it is as ordinary a part of reality as
is the notion that it is easier to walk through a door *after* we have
opened it. We are really quite used to using time as dimension; we do
it every day. We wouldn't be able to park the car if the cars that had
been there earlier hadn't moved on, would we? If last winter's snow
hadn't "moved" into the past, you'd still be shoveling, wouldn't you?
It explains how a person can be a baby in Boston and an adult in
Altoona. Scientists call the consideration for all four dimensions
"spacetime," which certainly sounds profound, but you and I should
think of it as rather normal, every day stuff.

Spacetime is supposed to depart from the ordinary in the presence of very strong gravity, and it seems to be so, because Einstein said so, and tests have shown that starlight does indeed bend as it passes near large bodies. This is one of the peculiarities that we have come to accept in spite of its peculiarity. But as I'll mention in a moment, there is credible rationale that instead of spacetime changing, and light beams being effected directly by gravity, the density of the light-carrying characteristics of space is increasing, causing the beam to slow and deflect as all beams do in denser mediums. This is not a hint that Einstein erred, but that again, the mathematics might be misinterpreted.

Einstein, by the way, was also an excellent writer in that his explanations for the average reader were cleverly understandable. He used a magic elevator to demonstrate, among other things, that light beams *had* to be effected by gravity (the above thoughts not withstanding.)

If we were in his magic elevator and it was accelerating rapidly upwards, a horizontal light-beam entering through a hole in one wall would hit the opposite at a slightly lower point. Since we experience the acceleration as "gravity," our conclusion is that gravity deflects light. And, by golly, so it does. Note, though, if light had an infinite speed, we would come to a different conclusion.

If we were to take his magic elevator and go down next to a source of very strong gravity: a black hole, say, and study its effects on the nearby space, what would we observe?

Nothing!

Relativity tells us that when we are in situations of distorted space we may not know that, because all of our measuring devices would be distorted also! That means to us that if one or more of the dimensions have been distorted, our ruler or measuring tape would have experienced the same distortion, and our measurements would be the same, regardless. To include time! So, based on what they tell us about relativity, it doesn't make any difference to us whether we are in a pinched spacetime situation or not. And if that is so, pinched space-time is meaningless, and ...a meaningless method of starting a universe.

On Gravity:

When I get on the scales I demonstrate that gravity does indeed work. Too well. Lifting a few bags of flour would convince anyone that gravity is a potent force, so much so that Einstein's deifying of gravity's role in the scheme of things was universally accepted as a given. The precise measures and predictions of solar system movements justify the perception that science has an absolute understanding of gravity; and who would think otherwise? It was a surprise, of course, when gravity was determined to be merely the "curvature of spacetime," but everyone seemed content with this *newer* absolute understanding of gravity.

Not everyone.

There is a great deal about gravity that still mystifies. In spite of its affinity for flour bags, it is shown over and again to be the weakest of the fundamental forces, such that even with the most exacting of efforts we still can't determine an accurate gravitational constant.[59] We do know that gravity is determined by $F=Gmm'/d^2$, where G is that inaccurate constant, and m and m' are the mutually attracted masses. But what is mass? Well, we say that mass is a certain number of atoms, or that the amount of the masses involved is determined by the force between them. We know that mass is related to energy by $e = mc^2$, but the amount of energy is dependent on the amount of mass. We know that mass is related to inertia, and the amount of either is dependent on the other, but that too is relative. Indeed, the values of mass, energy, and inertia are all relative and *by agreement*. Agreements or no, what is there about mass that space cannot exist without it? On top of all this, it has been demonstrated that mass increases with speeds approaching the speed of light, such that a bag of flour at the speed of light would have an infinite mass, an infinite inertia, and exhibit an infinite force.

Gravity may well be merely the curving of space by mass, but how does mass curve space? What is there about space that it is effected by mass? All that inherent energy? Energy has mass and gravitational influence, but are we saying that the gravitational effects of "empty" space are effecting mass? But space is ubiquitous, so any force would be self-canceling; and gravity is supposed to be a linear force, so why the curving? In the Meta universe, space is not curved but of varying density. The "light carrying medium" is denser near mass, and light is effected much like it is in the denser medium of water.

"...there are very few observations about gravity which cannot be adequately explained by Newtonian gravitation..."[235]

Gravity waves traveling at the speed of light, resulting from accelerating masses, are theorized. Gravity is more than merely a force between masses then, if it can radiate past the masses. Curvatures or densities can be visualized as radiating from a source – in waves – in a medium, but is this medium the same as a light-carrying medium? Gravity cannot be shielded, but what about gravity waves? If something is traveling though space, medium or no, it can be blocked or deflected. Not so for gravity? Gravity waves from that biggest of all accelerations of mass, the Big Bang, are credited with causing some of the observed ripples in the microwave background radiation, but the gravity wave would have been spherical and effected radiation equally everywhere, so maybe not. Vibrating cosmic strings are supposed to generate gravity waves, but none are detected. No gravity wave of any kind has been detected anywhere, in spite of repeated, intricate attempts.

And then there are gravitons: hypothetical bosons that carry the gravitational force which couples masses; suggested to be able to drift off on their own in outer space to collect in galaxies.[73] "Traveling ripples in the curvature of spacetime." "Might have to travel much faster than light." "The particle equivalent of gravitational waves." ...Well. This aspect of gravity deserves consideration also. According to careful measurements, gravity does indeed act faster than the speed of light. But if gravity is the result of faster than light gravitons, why is this not a violation of Einstein's tenet that nothing can do that? Does mass radiate gravitons? Do the gravitons go from only one mass to another, like an electron will go only from a cathode to an anode, or do they go both ways? Why would they reverse course? Do they collide? What propels them? How does a graviton "pull" one mass towards another? How can any particle, of any kind, pull anything? Also, why do we need gravitons if gravity is the result of space curving?

And then there are classical gravitons, (CG's) proposed Meta particles that have a maximum effective range of about a kiloparsec (3,000 light years), travel at around 10^{10} the speed of light, and *push* masses together.[125] This may not be as far-fetched as it seems, since hypothetical tachyons, whose speed is always greater then the speed of light, are theoretical considerations that might suggest a possible relation. The concept of particles pushing things sets better than particles pulling things, but CG's are not widely accepted and questions still remain. What propels a CG? Does it have mass? Can they, or any graviton, or any boson be disturbed by other force particles?

Good old, ordinary gravity remains an enigma. Yet scientists continue to apply dogmatic limitations to their efforts to quantify it into the Grand Unification Theory to describe the workings of the universe. All results so far indicate that the concept is invalid, and that it is utterly impossible to test it much further, but its theories have been used to explain and justify many other otherwise insupportable claims. Black holes, wormholes, Big Bangs and their kin will probably be with us until gravity is looked at again.

"...the final state of the system cannot be predicted with certainty if there is any error (no matter how small) in the measurement of the initial conditions."[98]

"...unable to subject their findings and theories to experimental scrutiny, they [particle physicists] have moved into a world governed entirely by mathematical and highly speculative theorizing."[95]

Chapter 16

Fundaments

* Waves Are The way * The Making of Matter * Hydrogen *
Molecules Are Different *

On Waves:

Gravity's possibly limited range would mean that the larger universe is
much like a gas in that particle (galaxy) motions and their so-called
structure are more effected by wave energy than anything else. The
sun's corona now reveals that low frequency waves might be whipping
oxygen atoms in such a frenzy that they indicate a temperature of 100
million Kelvins, which is about 100 times higher than expected.
Everything in nature is in the form of waves and wave patterns. Even
the basic particle is now referred to in quantum terms as a "probability
wave." Scientists studying vibrating particles on a vibrating plate
observe that those particles display all kind of voids and amplifications
that develop into sustainable, intricate, reproducible patterns; even piles
of particles called "oscillons" that approach each other, even dance
around as a unit, (if in opposite phase,) but avoid each other (if of the
same phase,) depending on frequency, resonance, and other natural
characteristics.[215]

Our weather is of wave nature. High and low pressure gradients,
temperature and humidity gradients, jet streams; ...and the patterns
flow around the Earth in waves of change. Ever so. Our weather, of
course, is caused principally by temperature changes that effect
pressure, and wind; and motions that effect temperature, ...pressure,
...ever so. Our weather is even more effected by the Sun: our star; our

very average star that acts like all similar stars in producing radiation that effects every thing around them. And all things around them suffer change: flows of some nature; in each case a type of wave action. Even barren asteroids experience pulsation's, expansions, contractions, quakes, undulations, throes and gyrations, in the so called "rhythms of the universe;" rhythms that are every bit in step with the wider-ranging rhythms of fluctuating energy fields.

On Basic Particle Production:

The wonderful term, fluctuating energy fields, is thrown at us frequently as something mystical: the manna of the cosmos, from which all good matter flows. And so it probably is. Not to beat a dead horse, the mathematics that excused the Big Bang concept is not erroneous, but merely erroneously applied. Instead of one large orgasm of short duration, a truer interpretation is one of a pleasurable, continuing, after the fact tranquillity. The two quandaries for the cosmologists have been the redshift and the manufacture of hydrogen: redshift is fully discussed; hydrogen manufacture is about to be.

Remembering that we popularly account for almost all element manufacture in supernovae, still is left the creation of hydrogen and helium that supposedly couldn't be produced by anything other than the extraordinary temperatures that are envisioned. But we addressed that concern with the realization that even those basic elements were finally made at normal stellar interior temperatures. It wasn't so much hydrogen atoms, perhaps, or even their protons that were the issue, so much as was the creation of quarks and their sub-substance siblings. And if empty space *is* brimming with energy on the order of 10^{80} ergs per cubic centimeter, then the requisites of the fusion process and the making of quarks are available, ...even in empty space.

But for those fluctuating manna fields, the production of sub-particles is far less demanding than the production of hydrogen, or galaxies, ...or universes. You and I witness changes of state daily. Everything from turning water to steam or ice, the condensation of clouds to rain or snow, the interaction of the solar wind and the polar magnetic fields for the display of the aurora borealis; even a window's catalytic conversion of moisture into a frosted pallet. Yes, yes, of course there is a difference, but changes of state is the determinant; the ongoing universal determinant.

Equations attribute energy fields with an average value of zero, but it is recognized that an average doesn't rule out a localized value for energy in one place being equal and opposite to an energy value someplace else,[225] and fluctuations, with the right coercion, can result in the birth of matter. Lots of it.

On Virtual Particles:

We are told that "quantum fluctuations" contribute to the "energy density of the vacuum," and virtual particles, in pairs of a particle and an antiparticle, constantly show up, interact, and "snuff themselves out." Now this is not as weird as one would think. There has to be a mechanism for making mass out of energy. A transition: that change of state: one in which sub-particles or protons and neutrinos are the easiest to form. Empty space, after all, is not supposed to be empty; ("The vacuum is thick with short-lived virtual particles.") and all those fluctuating fields of energy are easily transitioned to particles.

It is said that pairs are constantly appearing and disappearing.[64] They wink in and out of existence in a display too feeble and quick for us to notice; but if there is an external source of energy apart from the vacuum, the virtual particles can get an energy boost that will permit them to remain in the observable universe.[92] We are told that the energy of virtual particles ought to warp space. The deformation would be independent of that wrought by ordinary matter, which is said to cause space to contract; but while virtual quarks and electrons cause space to contract just as do their real counterparts, virtual photons or other force-transmitting particles cause space to expand.[31]

Virtual particles are said to be strongly effected by gravitational fields, but the gravity would have to be present before the particle. Since all nearby virtual particles would be influenced by the same gravitational field, they would all be treated the same. Quantum theories are interpreted that virtual particles have to be born together in pairs, simultaneously, in the same spot. But unlike virtual particles born in the narrow constraints of a laboratory experiment, whatever situation prompted the birth of a particle in the vast reaches of space would produce a flood of similar particles. And, because of the far-reaching initial conditions, it is likely that one area would produce large amounts of matter only, while another, more distant area, would produce large amounts of only antimatter. The energy source for potential virtual particles *cannot* harbor the makings of both matter and antimatter, because it has to be one and only one type of energy, not two.

If m+ = matter, m− = antimatter, and $E = mc^2$, then $E / c^2 = m$. But, if m+ plus m− = 0, then we're in trouble.

A particular potential voltage, positive, say, can act differently on positive and negative particles, but the voltage itself is one, and only one potential. One entity; one identity. And, ...there is a difference between a voltage effecting particles and a source that creates them.

We said a moment ago that "quantum fluctuations" permit virtual particles to show up, interact, and "snuff themselves out." That same quantum physics that describes such virtual particle activity is said to be applicable to the birth and death of entire universes! Universes and anti-universes can *pop* into existence ...only to snuff themselves out![225] We are told that if there is an external (?) source of energy apart from the vacuum, the virtual particles (or universe) can get an energy boost so that they can remain in the observable ...universe (?)!! [92] "External" sources can permit universes to remain in the universe? ...*Yes, ...this is the standard thinking*! Now, ...we realize that imaginary numbers are important tools of the mathematician, but too much imagining can lead to a departure from reality. Quantum theory, of course, says that there is no such thing as reality, but creditability certainly suffers.

Quantum theories have projected that virtual particles, or universes, could have been born out of nothing, but that same mathematics might also apply to the birth of sub-stances, sub-particles, or hydrogen.

On Hydrogen:

A repetitive theme is that the constituents of hydrogen, deuterium, and most of helium could be made only in unimaginable temperatures, and that these simplest of components, then are to have been changed over the eons into every thing else. ...MAYBE NOT. If all matter was to have originated as hydrogen, for example, the early universe would have been nothing but a rapidly expanding cloud of hydrogen that, as we have shown, could not possible have contracted into anything at all. But, even the well-received notion of element manufacture in stellar interiors, and some of those items that are admittedly not understood, come back to bite.

In stars, it is understood that the various elements are manufactured at different, concentric layers, in their localized, markedly different conditions, such that one layer might be comprised of nothing but silicon, for example, and the next outer layer might be nothing but magnesium, then neon, carbon, helium ...and hydrogen. Other stars might have other arrangements at different stages. If oxygen, ...for example, is residing in a particular layer, does that mean that the star has finished making that substance? If it is still making oxygen, do the necessary energies and circulations disturb other layers of other substances? Aren't stars in constant agitation and turbulence? If a star is done making a particular material, is it done making all? If a star is done making something, is the star starting to cool? Also, of those stars where the distinct elements were segregated into huge secluded reservoirs, what happens when the star goes supernova?

Supernova detonations are the most ferocious of the natural universe. What *does* happen to that stratum of conspicuous dissimilarity in such a cataclysmic convulsion? *Could* the elements maintain their individuality? If they don't in a chemist's crucible, how can they in such a caldron? And, if they do mix, ...why do we see all kinds of magnificent, large – thousands of light-years large – clouds of nothing but hydrogen, nothing but oxygen, nothing but sulfur, et al – near each other, but separated? Not intermixed; not intermixing. Why? Should we add this to the list of those items that are admittedly not understood?

It is an idea accepted by almost no one, but couldn't those fluctuating energy fields; fields that are capable of generating stars, galaxies, ...and quasars, be capable of generating individual elements? If not, how does a large cloud of hydrogen find itself nestled up against a similarly large cloud of oxygen? You would think that an energy field or 10^{80} ergs per cubic centimeter could indeed, periodically, gush out particular, discrete elements depending on the particular degree of gush; every bit as does that process we call fusion. The acknowledged birth place for heavier elements, supernovae, aren't credited with *that* much energy.

On Molecules:

It is one thing to have clouds of elements, such as arsenic, selenium, hydrogen, oxygen, sulfur, silicon, magnesium, iron aluminum, ...enriched isotopes and others floating around the universe in unexplained ways, but molecules? Carbon dioxide, carbon monoxide, methane, silicon oxide, ...water vapor, ...deuterium? Probably made in supernovae? There is detected in the stratosphere, even *organic* molecules in the interplanetary dust.[104] Is all of this supposed to have been made in supernovae? But wait, only a few supernovae happen in any one galaxy in any one century. If there were great distances between those infrequent supernovae, how can their effluent, which is every where, be so universally spread? Especially, since most of it is supposed to have been used up in the formation of the stars and galaxies that we see today?

As we will be discussing shortly, the possibility of simple "life forms" having reached Earth, to even thrive here, is actually viable. The idea is incredibly preposterous, except for the realization that molecules, and organic molecules, seem to form in space in prolificacy; that those questionable forms of life seem to suffer an unrelenting compulsion for increased complexity; that even the simplest crystals exhibit attributes that we normally assign to only living things.

Guy Murchie showed us that simple, natural turbulence, the cyclic and pervasive vibrations that are part of all natural scenes, tend to promote the beginning of *life* even from material not usually thought of as alive. Per Bak's *How Nature Works* tells us that many living and non-living processes interact intimately with their environment, that there is an ever-onward self-consistent integration of the simpler into an ever more complex environment. And, ...and, Murchie's *Seven Mysteries of Life* shows us that because there is no scientific definition of life, or non-life, the entire universe might be thought of as alive!

Now, I am *not* claiming that the universe is alive, and neither is Murchie. What is being pointed out here is that the same grand Order that dictates the stirrings of stars, the hemorrhaging of plasma from universe-wide circulation systems, the seemingly sexual attraction of electrons for protons; that overwhelming compulsion for merely questionable life forms to be part of more complex, grander, livelier beings; has managed to assign to each and every universal bit and piece a contributory role in Our grand, living, ongoing scheme. For such a grand Order, the making of protons and hydrogen atoms is no challenge.

Chapter 17

Variants of the Big Bang Scenario

* The Other Versions Of The Big Bang: Curved, Closed, Open, Flat *
Entropy, The Player * Other Universes *

The Big Bang is certainly not a singular description of universe development, and the different versions need to be discussed if we are to argue that *none* of them are viable.

Starting with those in Chapter 3:

The Standard Model Universe:

The original Big Bang model was fraught with many unsolvable, deflating, and discreditable problems.

The Inflationary Model Universe:

THE Big Bang model was to have corrected most of the Standard problems, but the age problem wouldn't go away.

The Loitering Universe:

Invented because the Inflationary model, which "corrected" the Standard model, needed correction. Generally considered ridiculous.

Following, now, are disparate variations of that problematical Big Bang scenario, all of which might pertain to each of the above, and none of which has more credence than the others. Years of analysis, juggling, and hopefulness have made none of them more viable.

The Curved Universe:

This is the one in which the universe is equated to the surface of a beach ball, on which if you travel in a straight line, you wind up back at the original spot. A light beam in space is seen to be deflected or curved by gravitational masses, and if we can imagine enough successive defections, we can also imagine that light beam being bent completely around until it hits that original starting point; much like Magellan did, from the opposite direction. This is the one in which the astronomer, looking through the world's most powerful telescope, sees ...the back of his own head! Because of that, advocates say that space is curved. It could be argued, though, that there is no sense in building bigger and bigger telescopes; all we have to do is turn around!

Einstein said that the universe is curved, so we assume that it is. But, what in the world does that mean? Can a light beam go straight across the universe and come back to the same point? Without turning? We are told that a satellite actually goes straight through space, but the curvature of that space as effected by the Earth (in our case) causes the satellite to go into circular orbit. Is this merely a play on words? Einstein didn't think so, nor do most scientists, but what meaning does this have for us? The predicted, and detected, bending of light around the Sun is held to be proof that the universe is indeed curved, but when we see light and sound bent around denser parts of the ocean, we do not say that the ocean is curved. When radio waves bounce off the ionosphere, repeatedly, so that they travel around the world, we don't say that the atmosphere is curved. The bending of light around one mass can very well be bent the other way around another, so what does any of this prove? Whether the universe is curved or not, the question relates to a perspective, a way of looking at things, and not necessarily an absolute.

The Finite Universe:

This universe is so-called, because the beach ball has finite size, and that the light beam would have traveled a finite distance to the back of the astronomer's head. The trouble with this, and all such attempts to relate cosmos characteristics to real world experiences, is that most of such efforts *are* merely a play on words. Is the universe to be pictured as a hollow ball? Solid? *Is* there a relationship between the "shape" of the universe and Earthly physical objects? Of course not, but cosmologists have to offer "simple" explanations to we simpler types, and that is understandable. But, the scientists themselves start thinking in these terms because *they* can't picture the obscure characteristics that they themselves have ascribed to the universe, and it distorts not only their language, but the very way that they plan their research and interpret its results.

The Unbounded Universe:

This, because there are no boundaries on beach balls!

The False Vacuum Universe:

False vacuum: "A bizarre state of matter of extraordinarily high energy density that can split off from our universe (or another) to form its own distinct universe, separate from ours (or the other.)" False vacuums create themselves and other universes! There is no way that we can discuss this sort of thing in a meaningful way, because it is another of those privileged mathematical exercises that are beyond most of us. But, the beauty of mathematics aside, we can certainly reason that the concept of false vacuums seems terribly contrived. They certainly are not mentioned in any other discipline or endeavor, or in any scientific encyclopedia that I've referred to. Are they only applicable to the starting of universes?

Closed or Open?

The Big Bang model has an energy density problem. If energy density exceeds a certain value, the universe is said to be Closed. If the energy density is less than the critical value, the universe is said to be Open. And, ...if the energy density is exactly equal to the critical value, the universe is said to be Flat. Astronomers have no idea which of these is the true picture, or even whether the energy density is positive, negative, or zero; and this is called the flatness problem. There is perhaps no subject that drives cosmologists bonkers more than that of the closed or open nature of the universe. They simply cannot determine the ratio (Omega) of the true average mass density of the universe to the critical density needed to just barely close the cosmos.

If evidence means anything at all, and it is supposed to, then the universe is clearly open and will expand forever into a slow, chilly death. This is because the redshift measurements of expansion exceed the gravitational restraints of the observable matter. Hubble's Constant is supposed to determine the rate of expansion of the universe, and whether or not it's closed, flat, or open, but his cosmological test is fraught with complications, and none of the very best efforts to pin this value down have eliminated any of the contenders. "If the open universe we see today is extrapolated back near the beginning, the ratio of the actual density of matter in the universe to the critical density, might differ from unity by just one part in 10^{59}. [a statistical impossibility]. Any larger deviation would result in a universe already collapsed on itself or already dissipated."[270]

The Closed Universe:

In this scenario, the universe's own gravity will stop the expansion and cause it to reverse and pile up again in the tiny point from which it was said to have started, perhaps to expand again, and again. Because there doesn't seem to be enough visible matter for this to happen, however, and because a variety of theories demand "closure," invisible dark matter, of up to 300 times more mass than we can see, is conjectured to assist in the general collapse. This type of space is said to be positively curved. This is an unpopular concept. This model, though, seems the best for a pertinent discussion of *the law of entropy.*

Entropy is the natural increase of *disorder.* Entropy, and disorder, *must* increase with time in all natural systems. Undeniably, for entropy to naturally increase in a Closed universe, it must have been at its lowest at the time of the Big Bang, higher today, more so at the later stop-and-contract point, and higher still as it crunches back into that same tiny beginning, and it must be at its very highest at the end; ..in spite of the dictum of this scenario, ..that the beginning and ending points are necessarily identical in design. Here, nonsense ensues. Those points of mandatory sameness cannot have both the highest and lowest entropy, and the Closed universe concept (which is just as meaningful as the others, they tell us) falls apart. The law of entropy, venerated on Earth at least, demands then, that the Closed model cannot be; and since the Closed is thought to be just as viable as the others are, this weakness may reflect on them also.

The Open Universe:

This has it that the universe will continue to expand forever because there doesn't seem to be enough mass to "close it up." This type of universe is said to be negatively curved and is also very unpopular, even though latest observations, using Ia supernovae as celestial yardsticks, evinces the open model.[263] But, using supernovae as "intrinsically similarly bright" standard candles is counter-intuitive. Since supernovae are exploded stars of a necessarily transient, temporary intensity, their very nature would seem to negate any valid comparison of their brightness; since they certainly didn't all detonate simultaneously, the "when" of the observation must dictate the apparent brightness. We'll discuss this idea some more in the next chapter.

And, visiting entropy once more, ...one mathematician, Roger Penrose, argues "that the entropy of a black hole is vastly greater than that of a spatially homogeneous arrangement of the same amount of matter."[266] However, a black hole is the closest thing that we have to that Big Bang moment when entropy was to have been at its lowest; and the far flung material in an open universe comes closest to Penrose's "spatially homogeneous arrangement of the same amount of matter," yet, this is when entropy *must* be its highest. Also, the temperature of the universe is said to have dropped from 10^{27} K, down to today's 2.73°, and is scheduled to ultimately reach 0°K when the Open universe dies its chilly death in 10^{69} years, again, when entropy *must* be at its highest. But, the Third Law of thermodynamics that tells us that "The entropy of ordered solids reaches zero at absolute zero."[276] The universe is not supposed to be a solid in its final stages, but the consideration seems applicable. So, here again, nonsense ensues, and, the law of entropy offers no support to the Open model either. No one likes either the Open or Closed versions, so the in-between version, the so-called Flat version, is the concept of choice. This too, though, has its problems.

The Flat Universe:

The Flat universe continues to expand forever but at a rate strongly influenced by gravitational forces. The expansion gradually slows down over more billions of years and finally almost stops. For this to happen, the universe would have to be at critical density, but since no-where-near-enough mass is visible, dark matter, inevitably, is needed here, also, to make this model work. This type of universe is said to have no curvature. A critical density, a very, very, very critical density, would be required to just balance the expansion with gravitation. The trouble is, that the required balance of forces is so exact, that the chance of it happening would have to be something like one in a thousand trillion, and no measurements, or mathematics, or even theory supports a concept of such exactness. "It would take an enormous amount of luck" for a Flat universe to evolve, and it is just about mathematically impossible.

As we said, scientists favor this model, even though there is no scientific justification whatsoever for their choosing this over any other. Why is this idea popular? Well, if you and I were given the choice of a universe scheduled for a slow death, one scheduled to collapse in a big crunch, or a universe scheduled to go on forever, which would we choose? We all, scientist and not, consider an on-going Flat universe far more palatable. It's merely intuitive, of course, but scientists are human also. It should not be missed that the Flat, on-going universe, the one that is almost mathematically impossible, is the closest to an infinitely lasting universe that *could not* have been born in a Big Bang, and the closest to what we observe!

The Other Universes:

A very popular idea that seems to be shared by most scientists is that there are other universes out there; *many* other universes, perhaps an infinite number. This concept stems from a number of different considerations. The Big Bang itself, for example, is thought to have been the result of a pinch of another universe's spacetime! Through a wormhole! And most mathematical equations said to depict the universe, have so many derivatives, determinants, fields, sets, variables, and solutions, that if reality and meaning are to be attributed to any of this math, a many-universe model is almost mandatory.

But this of course, is merely conjecture. There is no test – not even a thought process – that can lend credulity to the idea. We have no valid idea what our own universe is or looks like, so speculation about others is mere mental meandering. Does it have meaning to suggest that there is more than one universe, perhaps an infinite number of universes, instead of them all being part of the same one? A concept of many individual universes as opposed to many parts of the same one would demand that they be measurably different. Since size and shapes would not help there, the only way that these universes could differ would be in their natural and physical laws. Are the laws different there? Are their protons heavier? Unlikely, since our universe supposedly came from theirs, but if this is the case, what happens when universes with different natural and physical laws bump into each other? The mathematical derivatives and solutions ignore such realistic interpretations. If you add up all the other universes and call them one, then this universe couldn't possibly be a pinch from another.

The above are the more frequently discussed fundamental versions of Big Bang universes. The Flat is the universe of choice for almost all cosmologists, but no matter how hard they try, they can generate no more evidence for the Flat than the others.

The Anthropic Universe:

There is at least one more concept of the universe that is favored by some; the so-called Anthropic Principle; and, ...there are two popular versions; the Weak, and the Strong. The Weak Anthropic Principle has it that the universe evolved *for the purpose of supporting carbon-based life forms,* while the Strong Anthropic Principle posits that the universe was designed *for human beings.* Such conclusions are seemingly not within the scope of a science book, because they relate to the *"Why?"* of it all. Clarence Bennett, beloved old physics professor at University of Maine, used to tell us: "Ask me *how?* – Don't ask me *why?*" A nice feature of an Anthropic universe, though, is that we can deem the values of the natural constants to be *whatever we want them* to be merely because we are here to do such deeming.[225]

> "One possible explanation, [they] and other well-respected physicists argue, is that the universe was designed."
> Certainly more and more top-level scientists are considering the Anthropic Principle seriously in their work.[282]

This is a fine concept, but not scientific.

> David Lingley's concern is that the field might "..retreat from the high ground of science, becoming instead a modern mythology. This would surely mean the end of physics, as we know it.[95]

> "...if the universe could create itself, it would embody the powers of a creator, and we should be forced to conclude that the universe itself is a God."
>
> *The Seven Mysteries of Life*

Chapter 18

Latest

---

* *Ia Supernovae Do Not Anti-Gravity Make* *

The recent observations of distant Ia supernovae lead to the exciting certaintude[291] that the outer universe is experiencing even anti-gravity. Let's look at that.

First:

As was said, stars at least 10 times the size of our Sun are thought not to die quietly, but spectacularly in monstrous explosions called supernovae, that usually leave extraordinarily dense (10 to 12 miles across) neutron stars as remnants. Normal stars about Sun size, though, are said to have their nuclear fires die out gradually, quietly; allowing the star to slowly, ultimately, shrink into merely glowing ever cooling, Earth-size cinders that are called white dwarfs.

Ia supernovae differ from the ordinary in that they are said to result from two of those normal stars, each perhaps two or three times bigger than the Sun, that find themselves closely orbiting around each other and trading material until one collects enough from its neighbor to become denser and denser until its core explodes in a nuclear blast.

Ia's are confidently used as standard-candle yardsticks of distance, because a relationship has been found between the brightness of an Ia and the time it takes to expand to a certain size. The relationship of its size, brightness and expansion-rate implies an intrinsic brightness, and satisfies the criteria of standard candles. Distance, as was said, can be inferred from the observation of such standard yardsticks, and Ia's are used even in deepest space.

Well, now.

Recent observations are that the more distant, more redshifted Ia's seem to be *dimmer than expected*, which suggests:

> That they are farther away than their redshift calls for, and...
> That the distant universe is expanding faster than the nearer universe And...
> That the universe is open, and...
> The cosmological constant might decrease with time and vary with location, and...
> That a new form of energy called *quintessence*, which adds a cosmic energy density that varies with time and space is needed.[274]   and...
> That antigravity must be at work.[292]

Discussion:

So, there you have it, Ia's at a given redshift are dimmer than expected, so new forms of energy, to include antigravity "must" be the explanation. To be fair, the invoking of quintessence and antigravity was not made without a lot of thought. Conclusions weren't just jumped to, but things are not at all improving in cosmology. In spite of a growing realization that something is wrong, each misunderstanding results in more and more of the exotic imaginings of Chapter 13.

> "At the very least the expansion is not decelerating as rapidly as once thought. Either scientists must reconcile themselves to kooky energy, or they must abandon or modify inflation."[292]

It seems that cosmologists never go back to reconsider some of the assumptions, estimations, or rationale that continue to force upon them bizarre solutions, ..so it is left to us. Consider:

1. We have shown, quite adequately I think, that redshift is an unreliable determinant. (I have references to at least 50 examples of such evidence in my drawer.) The proper employment of the scientific method or even the cautious hesitancy of reasonable researchers should preclude fantastic answers except as a last resort. Even Big Bang supporters have suspected that redshift-distance ratios might not work at great distances because galaxies differ in their properties, especially

over time, and the "pace of expansion has changed over time or the intervening space is warped"

2. Ia's are said to result when the compacted residue of two stars compresses enough that a nuclear detonation occurs. But the force of such an explosion, and its brightness, should depend at least on the type and amount of material in the core. Stars are not at all said to have the same material; are plutonium explosions of the same brightness as those of uranium? The sum amount of material involved here, by definition, varies from four to six times that in our Sun. The amount of material most certainly influences the degree, intensity, and brightness of the explosion, so the larger amount should result in a 50% greater event.

3. Even though the frequency of such happenings might be questioned, the scenario of two stars trading material in a death-spiral can't be discounted, because a few examples of something like this have been seen. But, two bodies in such an ever-decreasing separation must develop an ever-increasing spin-rate, and the inevitable centrifugal force would demand a rather gradual integration. The not-at-all suddenly, merely incrementally increasing compression in this scenario, would inevitably result in incomplete "burns" of varying efficiencies and degrees of uniformity. Partial, or even directional effects can be realized, as weapons makers know all too well. Indeed, the final ball of residue probably could not compact sufficiently until the spinning stopped! Once that happened, the *scenario* of making that now bigger dead star is irrelevant! And, what we have then, is merely a ball of residue that could be expected from the death of a normal star four to six times the size of the Sun. But, since supernovae are ordinarily attributed only to stars at least 10 times bigger than ours, enough doubt exists that we should hesitate to imagine exotic products.

So, the type and amount of material and the degree of its compression can vary considerably, as certainly can the timing of the event, suggesting that only very cautious interpretation should be made of inferred measurements. It seems reasonable that we cannot, with confidence, assume that Ia supernovae have the same intrinsic brightness, that their development could be as suggested, or that any of the resulting determinations are valid.

4. How long does it take to make a normal dead star of the type from which Ia supernovae are to be born? Our normal Sun, said to be about five billion years old, is in fine fettle, doing quite well, thank you, and popular estimates are that it will take another five billion years, for a total of 10, before it dies and becomes a white dwarf such as the dead star material being discussed here. If the universe is only about 10

billion years old, or so (the figure used in the subject reference,) Ia supernovae then *must* occur *only* in the first-generation stars, in that "special" generation that *must* have formed before galaxies did.[223]

We are told over and again, that when we look at the most distant stars, we are looking back in time to the earliest years of the Big Bang. Stars at those distances, they say, *are* those born right after the big event. Ia's at those 10 billion light-year distances are also of that era, and must be at least 10 billion years old. But, since Ia's are said to *result* from dead stars *already* 10 billion years old or more, those dead stars had to have been born 20 billion years ago, and yet another age discrepancy evinces nonsense.

5. The cosmological constant was to have had its value set at Planck time, 10^{-43} seconds into the big event, but these newer observations have prompted a suggestion that the cosmological constant might change depending on where and when it is measured. *The constant isn't necessarily constant!* These same observations suggest quintessence, which adds a cosmic energy density that also varies with time and space, but since it is a new invention, no mention of where or when it was born has yet been offered.

They tell us that gravity was born at Planck time, along with everything else; dark matter, anti-matter, quarks and there kin. So, it follows that antigravity *must* have been born at the same time, in that same place, along with gravity!

Wow! Not only must we question how all of those other opposites could have formed together, now we must determine how gravity and anti-gravity could form in that same wondrous coincidence. Don't they add up to zero? Gravity exists *only* (apparently) in the nearer universe, while, as we are told resolutely, anti-gravity operates *only* at the outer fringes. How or why did they separate? *Could* they? Gravity is supposed to have an infinite range! Doesn't anti-gravity? And, before anti-gravity was invoked, dark matter equaling 100 or 300 times visible matter was deemed crucial to the big Bang scenario, because there was not enough gravity to allow the Big Bang to work. *Now what?*

6.The newly envisioned increasing expansion rate means that the universe is *open* and subject to the entropy considerations of the last chapter.

7. Last, but not at all least, ..how can stars gravitate towards each other; how can their residue compact; how can stars even form or stay intact at all, ...in an antigravity environment? Nonsense ensues, again.

Chapter 19

In Summary

---

* A wrap up * An Appeal For Good Science * What We Still Don't Understand * On The Ease of Argument * On Your Judgment *

What is left unsaid? We have outlined a sufficient number of arguments and rationales to present the Big Bang for what it is, an amazingly illogical scenario. Logic is not foreign to science, and it isn't supposed to be. Every scientist inevitably, perhaps unintentionally, but inescapably puts every new observation into the context of his or her own experience. Just like unusable solutions to complex equations are simply discarded, observations or theories that make no sense are routinely shrugged off and ignored. And they should be. Here are a few thoughts.

On The Illogical:

That there could be both hot and cold dark matter, especially together.
That dark matter particles could be as large as planets.
That those particles can intermingle with themselves but not galaxies.
That dark matter could be 300 times more plentiful than regular,
    real matter.
That the universe was ever hotter.
That the gravity constant could change with time.
That the cosmological constant could change with time and place.
That black holes could form at 7% of the universe's age.
That very specifically described homogeneities and inhomogeneities
    survived all that trauma.
That there could be Great Attractors.
That space outside a galaxy can expand, but not inside.

That space, or time didn't exist before.
That density could be the same everywhere in a supposedly spherically
   expanding universe.
That there are exactly 26 (or 10, remember?) other universes, from
   which ours was born.
That we could ignore relativity in that first instant.

There is no end to this list. There is almost nothing at all in the Big
Bang scenario that doesn't offend logic. Granted, logic is not a major
determinant, but it is one. One of many.

On Redshift:

NASA was experiencing problems with a number of its satellites in
different parts of the solar system; Pioneer 10 and 11, the Ulysses
probe, and Galileo have all been experiencing more gravity than they
should, as indicated by the redshift of their returning signals, since as
long ago as 1972. "Some scientists wonder whether our theories of
gravity are wrong." "We've accounted for everything that we can think
of."[285] [That redshift could be in error was not suggested.]

On Dust:

Dust. Recognized only as bother; a hindrance to efforts seeking the
"important" displays, but it is a significant player. We have all seen
glorious pictures of dazzling spectacles of celestial wonder that have
been partially obscured by almost solid clouds of that darned stuff, such
that we seemed compelled to study the gaudy while overlooking the
duller; just as we adore the flowers but ignore the leaves.
     It is every where in the universe, ever in the way. Blamed by some,
as the true source of the microwave background radiation – but it
doesn't radiate, so, no one speculates on its ingredients or age. It has
been seen associated with the very "oldest" galaxies, in the very
"earliest" parts of the universe; and, if so, it had to have been made
well before them. We don't really know what is. It surely is not atoms,
but molecules of some nature; molecules that seemingly had to have
been born in "long-ago-gone" earlier supernovae. There is so much of
it, that one heck of a lot of those "infrequent" supernovae would have
been required to produce it all, which compounds the challenge for
supernovae. The staggering, profuse ubiquitousness of just plain old
dust is evidence enough, by itself, that Big Bang concepts of ages and
creation are wrong.

On Antimatter:

Yes, we *have* pointed out a couple of times now, that the most acclaimed, most celebrated, most favored ingredients of the Big Bang ball are all opposites with each other. Hot and cold dark matter, homogeneties and not, and matter and its anti. Mathematics or no, we argued that opposites of anything cannot be made together. If this mathematical exercise has been misapplied, then so can others be. It *bears* repeating. Whatever that positive force was that precipitated matter in one area had to have its negative precipitating antimatter some place else. Period. Probably at the same time, if that fits the equations better; but they did indeed happen in different locations. Period.

If so, then entire galaxies of opposite material may well be (and do seem to be) floating around the universe, bothering no one, until the occasional collision produces a "good show" or quasar – if that's the way quasars are made. Perhaps it is possible that a lot of the resulting gamma-rays simply don't reach us, ...or are transformed to x-rays! The microwave background radiation, remember, is supposed to be fatigued gamma-rays, and it is much easier for gamma-rays to become x-rays. And, ...there are detected, zillions of otherwise mysterious, unexplained x-ray signals constantly bathing the Earth.

On Black Holes:

Black holes. They are the darlings of the cosmological world. The explanation to every thing otherwise unexplainable. They are said to be all over the place, but the dozen or so that have been "detected" have offered precious little in the way of meaningful evidence. They may indeed exist, but while some argue that they must, others argue that they can't. They don't make much sense in the grander scheme of things, but as far as the Big Bang scenario goes, it really doesn't matter. Remember Roy's Rule?

"If it doesn't matter, forget it."

On Assumptions:

It seems that cosmologists feel quite comfortable in making assumptions as a basis for just about all of their "theories." Even in relativity, scientists must introduce liberal assumptions that greatly simplify Einstein's equations. But the assumptions relied upon in Big

Bang thinkology are profound: that initial conditions can be arbitrary, that temperatures can exceed 10^{27} K°, that you can exceed the speed of light in a singular instance, that the universe can start from nothing. The theory rests on too many untested, and in most cases intestable, assumptions. From nothing at all, a pinch of space-time, etc....all of them well out of reach of verification.

On Age:

Years ago, Fred Hoyle of the University of Wales, proposed that life didn't originate on Earth; that bacteria and viruses could easily survive entry, and that perhaps meteor showers could have caused some of the historic plagues. The suggestion was treated as preposterous, and seems so. The idea that bits of life could be floating around the solar system, or perhaps even into the cosmos, does indeed seem unlikely.

But recent evidence in some of the Earth's oldest rocks shows that living organisms evolved on Earth as much as 4 billion years ago, right in the middle of the hellish bombardment that inflicted the planet's inception. And, ...recently discovered, living two miles deep in the Pacific, is a one-cell creature, who's "DNA is like nothing else on Earth," thriving, still, in boiling temperatures, gulping iron and sulfur and belching natural gas; "significantly increasing the likelihood that life perhaps does exist on other planets."[221]

Well! None of this rules out the possibility that this first life did originate right here. But until visiting viruses are ruled out, the significance, for us, of simple life forms coming to Earth from deep space, along with the prolific abundance of plain old dust as we have discussed, is the incredibly, perhaps infinitely, long times involved, in *any* scenario of universe development.

1. The first hydrogen molecules had to have formed.
2. Gravity in the Big Bang, we are told, waited until the universe's density was very much lower than that of Earth's atmosphere before beginning to act on matter.
3. Gravity had to gather the now greatly disbursed particles, all of which were still subject to radial separation forces and gaseous pressures.
4. The hydrogen clouds had to condense, in supposedly halting stages, until the first stars could form.
5. Some of those stars had to continue to grow, supposedly creating all of today's observed elements, except for

hydrogen and perhaps lithium, until those stars reached supernova stage.

6. Each supernova had to detonate, spreading those elements helter-skelter throughout space.

7. Those elements had to somehow gather into another series of stars, such as ours, that later formed planets, such as ours, that later formed life, such as on our planet.

8. Then those other, earlier planets on which life had formed had to somehow explode and spread those life forms helter-skelter.

9. Those now space-borne alien life forms had to travel from there to here, that they might rain on a planet such as ours, perhaps as long ago as 4 billion years.

10. Those alien life forms had to learn to adapt to Earth's environment, and thrive, so that they could be discovered and reported.

Now, just how long should all of this, or any one stage of it, have taken? The possibility of such alien life is a debatable argument, but it does suggest a universe older than most of the other, sterile developments; and certainly, much older than anything a Big Bang could provide.

On The Cosmological Constant:

If Einstein had assigned a value of zero to the darned thing, he might have realized that the universe could be EITHER expanding OR contracting![234] The mathematics allows either, or perhaps BOTH! (Expanding in one part of the universe while contracting in another!) But, quantum physics and cold dark matter theories demand that a cosmological constant of zero would mean that the energy density of space is zero, and we know that space is chock-full of all kinds of radiant energy. *Quantum theory just may well be wrong in this case,* and perhaps should not be used in cosmology at all.

Handily, the Anthropic Principle might be invoked to resolve the inability to assign a value to the cosmological constant. The Anthropic Principle has it:

That if a constant must have a certain value in order to support the eventual development of life as we know it, then it cannot have any other value, or we wouldn't be here to theorize about it.

That our very existence justifies our deeming the values of natural constants to be *whatever we want them to be*, merely because we are here to do such deeming.

So, ...that non-gravity, non-electromagnetic, conceived force; not considered by any scientists other than cosmologists, but imperative that it supplies 80% of the universe's energy density in an effort to explain the microwave background radiation; has an absolutely indeterminable value, but it doesn't matter because we are here to theorize about it.[225]

On The Forming of Galaxies:

Cosmologists simply do not know how galaxies form. They have toyed with almost every imaginable scheme, but each has its problems. We learned that the popular shrinking cloud of particles is just too unlikely a mechanism. Even that old familiar crutch, the black hole, doesn't explain it well, and a continuing flow of odd density waves doesn't do it either. There are scenarios that work well, as we discussed in Chapter 11, and will in Chapter 21, but not at all in a Big Bang environment.

On Fundamentals:

They may not be so fundamental after all. This is not negativism; this is a creed by which all scientists should live, and the search for truer fundamentals is more meaningful than the reiteration of perceived, latest to be accepted, "facts."

On Other Scenarios:

That there are other scenarios testifies that the Big Bang might not be.

On Science:

Good science is the development of repeatable, verifiable experiments and observations. Good science is the minimal reliance on assumptions, and is almost totally uninfluenced by beliefs or opinions. Good science is "self-corrective" such that erroneous assumptions, beliefs and opinions are ultimately discarded by a diligent adherence to the "scientific method." What is real in physics is what works.

Classical physics deals with reality, but quantum mechanics, now a corner-stone of cosmology, denies the existence of *any* absolute reality. Cosmologists rely heavily on the concepts of quantum mechanics, most of which are unproven, most of which need not apply to any system bigger than an atom. Physicists investigate the universe in accordance with the laws of physics, while cosmologists are more interested in changing the laws of physics to fit their theories, preferring the more beautiful theory over the more rational. Vaguely familiar with the mathematics of general relativity or quantum mechanics, the cosmologists get used to using it as a given. They have come to accept the actuality of various bizarre, unproven concepts to the point that there is a fear that society might reject the "arrogance of science" for more religious answers.[187] Fred Hoyle warned us against the "danger of doctrine," but the advice seems to have been ignored.

"The constraint of experiment on the scientific imagination is growing weaker."[129]

"Cosmology has ...pushed into areas once thought to be the exclusive domain of religion or mythology."
"They cannot deduce a complete theory of the universe from observational evidence and self-evident principles alone, so their models are always infected by their own views on what constitutes a simple, pleasing and elegant theory."
"Cosmology may come to be ruled more by aesthetic preferences or prejudice than by the traditional principles of science."[130]

On Pathological Science:

John R. Huizenga, in his devastating *Cold Fusion; the Scientific Fiasco of the Century*,[120] shows that self-deception is a characteristic of *pathological* science, ...which was defined by Irving Langmuir as... *"The science of things that aren't so."* Huizenga quotes Langmuir's suggested six criteria of pathological science (which seem to be applicable to cosmology):

(1) The maximum effect that is observed is produced by a causative agent of barely detectable intensity,[a] and the magnitude of the effect is substantially independent of the cause.[b]

    a.   Dark Matter (Can't detect it at all.)
    b.   Inflation scenario is supposed to be independent of initial conditions.

(2) The effect is of a magnitude that remains close to the limit of delectability or, many measurements are necessary because of the very low statistical significance of the results.

    a.   Microwave background variations are a mere 30 millionths of a degree, and only a very few possible examples of dark matter lensing have been detected in spite of very extensive searches.

(3) There are claims of great accuracy.

    a.   Measurements of observed element ratios are deemed crucial.

(4) Fantastic theories contrary to experience are suggested:

    a.   The entire list of Inventions and Hypothetical Concepts of Chapter 13.

(5) Criticisms are met by *ad hoc* excuses thought up on the spur of the moment.

    a.   "Special types" of black holes, "odd" density waves, "localized" anti-gravity, etc.

(6) The ratio of supporters to critics rises to somewhere near 50% and then falls gradually to oblivion.

    a.   We're working on this one.

It is inferred that any one of the six criteria justifies the title: Pathological Science. Five out of six is significant.

On What We Don't Understand:

In spite of the dogmatic certainty that cosmologists have ascribed to the Big Bang model, its foundation is shaky indeed. Not only are we able to present this long, damning argument, those same cosmologists admit that there are important aspects to it that they simply don't understand.

What is matter? They have literally given up trying to find out.
Why does quantum mechanics apply to cosmology? Does it?
The gravitational constant.
The cosmologist constant.

Why do the fundamental constants of nature have these values? What
constant are we missing?
What is the source of all the x-rays.
Why is the speed of light a limit? Is it?
Why does mass curve space?
What propels a graviton? ...Or any boson?
What is energy?

We do not have a full understanding of galactic development, or
gamma-ray busters that emit energy of more than one quintillion suns,
or great attractors equivalent of 50 billion, billion times the mass of the
Sun. Or bigger! We don't understand quasars, "unknown mechanisms,"
brighter than 10^{15} Suns.

And this too, is merely a partial list.

Agatha Christie's great detective *Poirot* says that a doctor that lacks
doubt is not a doctor but an executioner. Well, things aren't quite that
bad in Cosmology, but there are enough unknowns that we should not
be lacking of doubt.

On Entropy:

Entropy, which is a *decrease* of energy, or an *increase* in disorder,
increases with time in the natural process, ...and the reverse is
considered impossible: entropy decreases do not occur in nature.[268] So,
regardless of the mechanism of universe development, and regardless
of whether or not it is expanding, entropy is said to be universally,
inevitably, increasing with time. It seems, though, a matter of opinion,
as to what constitutes a "higher" entropy, as was discussed in the Open
and Closed universes of the last chapter, and ...it seems that entropy
argues successfully against *all* Big Bang models of the universe.

On Order:

The most obvious natural characteristic of your world and mine is
that of *Order*: a set of physical, chemical, natural – but sometimes
mystical – laws that control every aspect of reality, both observed and
unobserved; this side of the universe or the other. Even mysteries of
the quantum microcosm are in accordance with these laws, in spite of
the uncertainties that we have been unable to fathom; and beyond
doubt, so are the unknowns of the far reaches of space and all of our
universe. And these laws must be part of one larger law that relates all
others into an harmonious one, so that each and every boson and
fermion is in complete agreement with one another; such that every

happening and fluctuation, every burp and supernova, is mutually supportive in a grand continuum.

The affinity that electrons have for protons, that one hydrogen atom has for another, that crystal molecules have for nestling into special nooks, that soap and water bubbles have for consummate shape, ..that eggs have for exquisite design; are in accordance with the orders of that larger law.

Even disorder is as prescribed. Imperfection leads to complication, complication leads to turbulence; and turbulence leads to oscillations, vibrations, waves and rhythm. Whether it is the rhythms of music, or waves on a beach; whether it is the tones and cadence of a stair-stepping electron finding it's ordained energy level; that systematic disorder leads to such wonders as: ice crystal sheets forming on our window as we watch, the humming of ice on a lake as it stretches and waves, and patterns of spider webs; the shape of trees, the movements of continents, planets and stars; the 11 year cyclic metabolism of the Sun.

This Sun and all others, every galaxy and quasar; all operate in accordance with that one larger law, as do their plasmas and gasses and atoms and photons – wherever they are.

Now, this is some law! What are the odds that this boundless, eternal, invariably immutable larger law, perhaps the only perfectly written law – ever – could result from the catastrophic accident of an exploding pinch of nothing? You can make this call as well as any scientist – and better – because you are the consumer: the jury: the one who is expected to buy all of this.

Life and death are a part of the Order prescribed by this grandest law, and in the larger sense they are not different. We "borrow" atoms and molecules from the Earth only to return them later that they might be used by someone else or some thing else; and the only change is in fleeting personalities or identities. The atoms in us were once in stars and some day they'll return. Since living and dying are part of this great, ubiquitous Order, perhaps they could be likened to one gene in a grand DNA molecule: inherited for passing on – and passing on relates back to our continuum – and continuum is a frequently use word for universe.

"Thus, it is perfectly obvious that only special states could possibly give rise to the immense degree of order that we see in the world."[230]

It follows that the concept of death for the universe itself has no meaning either, yet the Big Bang model demands that the universe die, dead, in either a slow decay or a big crunch, even though no discovery of science has so much as hinted that the larger law and it's decreed Order have a shelf-life. Might such a concept be regarded as inefficient at best, wasteful perhaps, and of poor design? Should we expect that this otherwise apparently perfect larger law; a law that orders electrons and protons and their siblings to live, and socialize, and court each other in accordance with properly supervised, time-honored procedures – for billions of years at the very least – to have shelf-life?

The future predicted by our cosmologists is not at all comforting or satisfying; the universe is either going to expand, and cool, and gradually decay and die; or it is going to collapse back into a big crunch that will end everything more spectacularly – all because of one apparent, possibly erroneous observation, that of the redshift. That damned redshift indicates an expansion of the universe; inferring that it had to have begun at one point; inferring that the point was a singularity, inferring, along with quantum theory, that physics could be ignored at any time and any place.

Scientists, by nature, are positive contributors, and they have tried their damndest to justify a scenario that would permit a flat universe to live, but that doggone cosmological constant, which is demanded by the Big Bang model, just won't cooperate. In spite of the evidence and mathematics that almost prohibit a flat universe, the flat universe has an aesthetic appeal to most astronomers. Why? Perhaps without their realizing it, it is so darn close to an ever-stable, unchanging, infinitely old universe, that they like the idea in spite of themselves.

On The Obvious:

Finally, ultimately, inevitably, the play is over, the coup de grace is administered, and the curtain falls. For the Big Bang, though, there was not one, but two coup de grace, delivered together as almost one.

The perfectly symmetrical ring formation in Supernova 1987A, and the indisputable gamma-ray signature of matter and antimatter in collision, right here in our good old Milky Way, put the nails into the Big Bang's coffin. Supporters have dodged every other pointing finger, but there will be no avoiding these.

Because gravity could not possibly have created the solenoid-like structure of Supernova 1987A, never mind what did, the Big Bang's proclamation that gravity is the explanation for all things celestial is shown to be clearly, obviously wrong. And, that one bit of antimatter that just happened to run into our galaxy, in full view of all who look, is proof-positive that the Big Bang concept is failed.

On the Ease of Argument:

The main evidence in support of the Big Bang, remember, was the observed expansion, the microwave back ground radiation, and the apparently observed ratios of light elements. Well, I believe that the arguments that we have presented are sound enough that you, the jury, the consumer, the final arbiter of those issues that count, might now be convinced that the Big Bang scenario is merely an incredible façade. In addition to much, much more evidence, we showed that the observed expansion may not be real, but that even if it is genuine it wouldn't demand an explosive beginning; we showed that the microwave background radiation is not evidence of a big Bang as much as it is that a Big Bang couldn't be; and we showed that the ratio of observed light element ratios is most likely the result of the natural on-going processes of a infinitely old, infinitely large universe.

The interest here, though, is that it was so easy to present such a lengthy and thorough argument. Each and every assumption, rationale, and aspect of this complex fabrication has been thoroughly decimated, completely negated, and it has been *easy*.

On Your Judgment:

We can only believe that cosmologists themselves do indeed see the folly of the Big Bang, but they simply haven't been able to find a way out of the mess. Hopefully, your judgment here today will help. ...And, it may. Those diligent scientists that still subscribe to the time-honored scientific method, those that have been over-ridden and ignored by the mainstream dogmatists, have been unable to bring sensible reasoning back to an earlier-methodical profession. They need some support. Perhaps we kids on the sidelines can say it: *The emperor wears no clothes!* If you remember, you were challenged to determine whether or not cosmologists have given us their best efforts: their best results, specifically having to do with the origin of the universe. You are asked to decide, based on logic and reason, whether or not the so-called Big Bang model, in which the universe was supposed to have begun in one huge explosion, is justified by the evidence.

What is your decision?

Chapter 20

Alternative Universes

---

* The steady State Universe * The Quasi Steady State Universe *
The Meta Universe * The Plasma Universe * The Three Forces *

Fortunately, there are other, more palatable alternatives to the Big Bang, each of which suggests a universe that is infinitely large, and infinitely old, with no beginning and no end, eliminating all age conflict concerns, and just about all of the so-called problems.

The Steady State Universe:

The Steady State model might be considered the opposite of the Big Bang scenario. There was no beginning for the Steady State universe, nor will there be an end. It has been here, and will be, for all eternity. Until Hubble "expanded" the universe with his redshift observation, this *was* the widely accepted model.

The Steady State universe *must* expand because a continuous, constant, spontaneous introduction of matter, (which is triggered by the presence of mass) fills the voids with new-forming stars that replace the old, and forces the stretching of space. The density of space is conserved, but not that of matter, and the steady facility for the appearance of new atoms gives space active physical properties: it is not stagnant; it has a vigorous predisposition for giving birth to matter. The formation of galaxies is continuous, ongoing, and not of only one era. It is argued that there shouldn't be any basic differences in a Steady State universe, no matter where you look, or when; that a Steady State universe should look the same everywhere. This concept is the so-called "strong symmetry" principle. We are told, though, that the universe just doesn't look the same everywhere; that in a Steady State universe there wouldn't have been the super-hot, super-dense, beginning, so there shouldn't be the microwave background radiation that we see today. And that the steady introduction of matter would violate the conservation of energy.

Well, ...as we detect more and more galaxies and structures with each new observation, everything *does* seem to look pretty much alike regardless of where we look. As far as the microwave background radiation is concerned, we have already shown that it is quite unlikely related to Big Bangs, and perhaps cannot be a factor. The conservation of matter argument is a little stronger, but that it is an issue here and not with the Big Bang itself seems incongruous.

And, ...the Steady State doesn't need dark matter, because if the universe is a trillion years old, there should be plenty of old, dead stars and such that can provide the extra "required" gravity.

The Quasi Steady State Universes:

The Quasi Steady State Concept (QSSC) – Would you believe it? – is a modification of the Steady State model. This also claims no beginning or end. But here, instead of a continuous influx of matter, there have been a series of "mini-bangs," violent activity in localized "creation fields" of different energies, occurring at erratic intervals over all of time at the heart of galaxies and around galaxy clusters spaced billions of light-years apart, continuously agitating and restructuring; strengthening and weakening; the universe.

The localized creation fields are suggested as explanation for some of the extraordinary, jet-like, otherwise unexplainable energy sources that are usually blamed on black holes. And instead of seven, or 10, or 20 billion years of Big Bang expansion, the universe has been expanding for some trillion years and would have accumulated plenty of dead, normal stuff that would serve as the "perhaps *not* required" dark matter. Its authors can explain the amount of early deuterium and other light elements, and claim that iron filings from supernovae could be radiating and radiating, better explaining the microwave background radiation and its temperature.

Proponents claim that the conservation of matter is not violated because the periodicity of material introduction supposedly permits a natural supply of energy into the system; but periodicity, by itself, can't explain away conservation of matter laws. And, ...that creation fields exist only in adequate strength in regions of high mass density, such as in large clusters of galaxies or the dense core of an individual galaxy, seems counter-intuitive. Instead of a preponderance of mass triggering creation, wouldn't the *lack* of mass in empty space be better? ...The creation of matter in galaxies will be discussed again.

The Meta Universe:

The Meta universe is not your average universe. Relying on deductive, rather than inductive reasoning, an entirely different type of universe is developed. Where a lot of scientists start off by showing what is wrong with other models, Meta pretty much starts from scratch, and pieces a universe together, molecule by molecule, as it goes along. And, it *is* different. Even relativity is given a new light. The deductive approach has given the Meta model a credibility that others lack.

The Meta universe puts not only relativity, but quantum physics as well, into a perspective based on classical physics. It explains that the constraints on the speed of light are determined by a "light carrying medium," (which is denser near mass) much like the speed of sound is constrained by a "sound carrying medium" (air, water). Space is not curved, but merely of varying density. Redshift is caused by light passage through this denser medium and not by expansion.

Faster than light is possible for some things, such as gravity, which is instantaneous, but of limited range. The limited range for gravity in a Meta universe is not a new conjecture in spite of the widely held belief that it is boundless; but remarkably, Meta even explains *why* gravity works! We all know *how* gravity works, but here, at least, at last, the WHY of gravity is offered. Meta's gravity is the instantaneous *pushing* action of Classical Gravitons (CG's) on mass. The limited range of gravity means that the larger universe is much like a gas in that particle (galaxy) motions and their so-called structure are more effected by wave energy then anything else.

It limits us to *five* dimensions; that of space, time, and scale, but each is infinite. The dimension of scale recognizes that there is no ultimate smallness nor an ultimate largeness. There is no such thing as a fundamental particle, and even the observable universe is merely some of the stuff of a larger something else.

Its dimension of scale, like a limited range for gravity, is not necessarily new, but it is long overdue for discussion. One cannot justify a basic particle, such as the quark, unless the question of its makeup is ignored, yet if it is matter, it must be made of still smaller things. Quarks and their siblings are merely the latest level of smallness. Others are waiting to be discovered, but because larger particle accelerators aren't going to be built for a while, quarks it is for now. At the other end of the scale, we are, and forever, stuck with our largest entity, the universe itself, but the question remains, what is the universe a part of?

Meta says that quasars, supposedly of unbelievable energy because their redshifts indicate that they are at great distance, are actually quite nearby, their redshifts are unrelated to expansion in any way, and they are the sources for most of the mysterious x-rays that pervade the sky. The microwave background radiation also is a local phenomena, indeed cannot be anything else. Hence, there is no expansion and there was no Big Bang. And no black holes. And, certainly, no dark matter. Abundances are those that are naturally brought about after an infinite time, the universe may not be expanding at all, and all forces result from the collision of particles.

The Plasma Universe:

Perhaps a lot of us think of plasma only as a form of blood, but a physicist thinks of it as an ionized gas that usually contains equal numbers of positive ions and electrons. It is fascinating stuff. Intensely studied in laboratory vacuums, plasma is seen to develop into, and be worked as; glowing electric currents, colorful fleeting filaments, cords and strings, swirling vortices, magnetic fields, jets, directional signals, toroids; you name it. Usually, intense magnetic forces and energies are involved.

(By the way, if plasma can develop into string-like entities, why couldn't "string" theory be more applicable to "ordinary" plasma development than imagined exotica? The abstract, antiseptic characteristics of mathematics are frequently ignored when mathematics is used to justify a theory; but the universal beauty of mathematics is that it can be applied to anything.)

Filaments and vortexes can form impulsively, and grow larger and stronger as they develop in magnetic fields that concentrate the plasma in a so-called self-supporting "pinch effect." The stronger the field, the more the pinch, and the stronger the pinch, the more the field, and so on. Electrons trapped in such a magnetic venturi can be expected to emit microwave radiation, and from even steady-state conditions, rushes and surges of whirlpools of magnetism can result from subtle increases of the fields.[69]

Outside the laboratory, you and I can experience plasma as the Aurora Borealis, smell it after a nearby lightning strike, and see it in Sun spots. Yes, plasmas do exist in the real world, only more so. Much more so.

Plasma energy levels that can be handled in the laboratory are insignificant compared to what naturally occurs in the atmosphere, in the solar system, and beyond; even in stars, galaxies, and throughout space, For instance, currents and fields exist throughout the solar system, and filaments over a hundred light-years long have been detected in our own galaxy. While a typical lightning strike might generate 100,000 amperes, the electric current thought to be pervading the Milky Way exceeds 3 billion, billion amperes.

We have all heard of cosmic rays, and we have discussed them earlier, but they are mysterious even to scientists. Most are thought to originate as heavy-element particles that almost constantly enter our atmosphere with tremendous force, usually breaking up into smaller and smaller particles as they come though the atmosphere, not to be noticed, of course, by the average looker-upper. A point here, though, is that they do have tremendous energy; cosmic rays can have ten billion, billion times the energy of an atom in the Sun's corona! ...A second point might be ...why?

The most likely explanation, perhaps the only conceivable one, is that these high energy particles, these very many, very high energy particles, are propelled by extraordinarily strong interstellar magnetic forces, much like the particles accelerated in our super-colliders. It's a cinch that mere interstellar drifting wouldn't give the particle anywhere near that level of energy. Perhaps the particles pick up more and more momentum as they pass the many huge magnetic sources in their path, much like artificial satellites are given boosts along their way by being directed past moons and planets to gain energy. Such reinforcing changes of direction, in addition to the occasional kick by supernova shock waves, could deflect and accelerate electrons and cosmic ray particles so that they would be not only fast indeed, but almost impossible to trace; perhaps adding to the myth of a curved universe.

And there are indeed many sources that contribute to a boundless, pervasive, ubiquitous sea of electromagnetic forces: clouds of hot hydrogen in the Milky Way, emissions from the galactic center, even magnetic stars in which heavy elements are often found. The familiar Crab Nebular is 100 times stronger than the Milky Way clouds of hot hydrogen, and a source in Cassiopeia is a million times stronger than that: Perhaps a million, billion, billion kilowatts. To top that, the electromagnetic radiation of the constellation Cygnus (Said to be due to a collision of galaxies.) is about a million, billion, billion, billion kilowatts.

It has long been established that galaxies have magnetic fields. The studying of polarized light that reaches us, of all things, results in quite convincing evidence that our own galaxy is much like a magnet; one that requires an enormous electric current: the equivalent of trillions of lightning strikes, and the development of those fields has been rather well accepted and seemingly understood for many years – up to now. Recent computer simulations seem to indicate that individual star magnetic fields shouldn't align the way that they do, so, inevitably, it is deemed that black holes must be at the center of galaxies to make their magnetic field work properly. Even black holes, though, can't account for a lot of the observation, but they *are* the darlings of cosmologists, and the pervasive "answer" to a lot of questions.

There is still more to learned even from our own Sun, which certainly has been studied far more than any other. Recent discoveries of corona temperatures in excess of 100 million Kelvin's are about 100 times greater than previously thought. They have a couple of good theories of waves, frequencies, and oscillations of charged particles to explain those high temperatures, but the point here is that the observation of nearby, familiar, natural electromagnetic physical processes that result in "surprisingly" high temperatures should not go unnoticed. Even the thoroughly studied have surprises for us.

The turbulent frothing and lathering of the seas of highly ionized particles and gasses, the pulsating, vibrating and snapping of magnetic lines, and the high-speed solar wind that bathes us daily, which is fastest at the Sun's poles by the way, reemphasizes the electrodynamic nature even of ordinary stars. Additionally, the rotation speed of the Sun is better explained by electromagnetic considerations than gravitational forces. And, ...even the central core of our own Earth is now found to be every bit the *electric motor*, rotating at its own slightly *faster* speed.

In recent observations of star formation, "not only does gas fall inward, but vast quantities of gas and dust stream outward" in narrow jets and "giant peanut-shaped bubbles" 100 times more massive than the Sun. Seen, are bright blobs of gas, likened to the beads on a necklace, traveling more than 100 miles per second; much too fast to have been caused by a gravitational collapse of a dark cloud. "They stretched for more than a light-year from the cloud while remaining impossibly narrow." Also seen were "vast and extraordinarily massive outpourings of gas around embryonic stars, ballooning out on opposite sides of the newly forming stars and stretching much farther out than the jets ...in bipolar outflows." [224]

Referring again to the Preface in the front of the book, Supernova 1987A *has to be* electromagnetic in nature. We are observing a characteristic toroidal shape! Gravity plays no meaningful role, and the Big Bang *has to be* wrong!

Galaxy centers are frequently observed to be shooting out jets of magnetic material. In some galaxies, jets, or "chimneys," of charged particles, or plasma, emitting microwave radiation and traveling at more than 1,000 miles per second extend out trillions of miles: the largest single objects in the universe. They can carry more energy than all the stars in the Milky Way produce in 100 million years. By any stretch of the imagination, gravitational forces cannot explain a fraction of such energy or results. So of course, this activity is invariably attributed to black holes, even though the concept of black holes – sink holes – spewing anything at all is dubious; and this old argument pales in comparison to a more reasonable, ordinary electromagnetic rationale. Disregarded entirely is the more classical, more reasonable electric dynamo model of galactic structure, and the possibility that galaxies could be *sources* of energy rather than sinkholes. If so, perhaps an explanation for the rotational speed of the spiral arms would not require dark matter. If galaxies are energy sources instead of sinks, perhaps then, there is a correlation with the creation fields of the Quasi-Steady State model.

A galaxy spinning in a magnetic field – magnetic fields that are everywhere in space – is every bit a huge generator. A better reason for the dynamic disk shape of a galaxy is electromagnetism, not black holes, not gravity! The currents flowing around the disk naturally result in forces up along the spin axis, every bit the same way that solenoids and electromagnets work in you own home.

It has long been thought that electric currents flowing in intergalactic space could connect one galaxy to another in ever larger magnetic fields; (recent indications are that the fields might even be stronger than predicted) and galaxies formed within a large, fast filament seem to maintain the speed and direction of that filament. Might not this better explain all that galactic streaming that perplexes scientists enough that "great attractors" had to be imagined?

Where do intense radio waves come from? There are all kinds of radio sources: galaxy interiors, sources like the Crab Nebula, hot hydrogen clouds and multiple particle collisions when clouds of gasses collide. It is said that a disturbance in the "fabric" of space allows galaxies to precipitate and develop into the so-called structures, such that all the movement and interaction results in radio waves. Perhaps though, most of such energy stems from the high velocity movement of free electrons through what has to be a rather universal general interstellar magnetic field. Indeed, it is determined that all of this radiation means that there must be energetic clouds that are churning rapidly and at once traveling at high speed all over the universe and in all directions. None of this can be explained by the Big Bang scenario.

Regarde:

$$\text{Newton's Law of Gravity} \quad F=G \; \frac{mm^1}{d^2}$$

$$\text{Coulomb's Law of Charges} \quad F=C \; \frac{qq^1}{d^2}$$

$$\text{The Law of Magnets} \quad F=K \; \frac{mm^1}{d^2}$$

Isn't there a striking similarity in these equations? You're familiar with the first: the Law of Gravitation. The others are the laws of electric and magnetic forces receptively. Three of the most fundamental laws of nature: simple, straight forward, elegant. Unrelated, but identical of form; amazingly similar, but with crucial differences.

Science, especially physics, historically attaches significance to form, and from it has developed most of the major theoretical advances. Reflecting on the evident equivalence and harmony of these three equations; even if we didn't know anything at all about them, we could intuitively expect them to be equally important, equal in effect.

No! Not so in this case.

There are two profound differences in the equations:

1. Electrical and magnetic forces can be of attraction *or* repulsion; gravity cannot.
2. In any corresponding situation, the last two exceed the force of gravity *by 40 orders of magnitude!* To put that in perspective, one source has it, that 40 orders of magnitude is about *the difference in size of an electron and the size of the entire observed universe*!

These two statements are of crucial realization and perspective, yet almost everything in modern cosmology ignores them. By far the very weakest force in the universe, gravity is deemed to be the most important, in spite of the fact that esoteric, exotic, alien and bizarre hypothetical crutches of imagination (strings, dark matter, black holes, etc.) are necessary to defend it.

The plasma model of the universe does not consider gravity to be the dominant force. It postulates rather, that most of the matter in the universe is in the form of ionized gas and that electric and magnetic fields are the determinants; that the Big Bang never happened, and the universe is infinitely old. Also, microwave background radiation might be a scattering of ambient radiation by small filaments of plasma, having nothing at all to do with a Big Bang scenario!

Well, there you have it. Not only have we been able to show that the Big Bang could not be, but that there are indeed other, more satisfying possible scenarios. If you have made your decision, fine, but surely the job is not done. Discrediting the Big Bang is quite an important accomplishment, but that leaves us with an empty, nagging question to answer:

What the heck *is* the universe like?

Chapter 21

"Our" Universe

A Straight-Forward, Useful, Compilation of Characteristics that Comprise a Conjectured but Palatable Universe

If Not The Big Bang, What?

Yes, Virginia, there are those other models; the Steady State, Quasi Steady State, Meta, and Plasma are the versions that we have presented as reasonable alternatives to ...that other one.   And they really are worthy alternatives. Each could be accepted almost without reservation; each has a number of meaningful characteristics to offer and for us to discuss:

> That there was no beginning, and will be no end to the universe: on that, all these models agree.

> The universe is expanding:
>> The Steady State model must expand, forever.
>> The QSSC has been expanding for a trillion years.
>> The Meta and Plasma aren't.

> The microwave background radiation is explained by QSSC, Plasma, and Meta, as a normal, perhaps local characteristic.

> The Temperature of the MBR is explained by QSSC.

> Continuous creation is demanded by QSSC.

Relativity, the speed of light, and quantum mechanics are explained by Meta.

Space is not curved in a Meta universe. The "light carrying medium" is denser near mass, and light is effected much like it is in the denser medium of water.

Redshifts, as explained by Meta, are caused by light passage through this denser medium and not by expansion.

Gravity is explained by Meta.

String theory is representative of, and cosmic rays and star and galaxy formation are explained by, electromagnetic forces and the Plasma model.

Meta gives us the dimension of scale: there is no ultimate smallness nor an ultimate largeness.

Elementary particle abundance is explained by QSSC and Meta as naturally brought about after an infinite time:

Star birth parallels galaxy birth.

Quasars are neophyte galaxies.

Black holes are irrelevant or do not exist.

Structures are merely the result of wave action if they exist at all.

There is one, and only universe.

Well, now. Even among the more viable alternatives to the Big Bang, minor differences do exist. A happy note, though, is that none of the alternatives require exotic invention, and each of them is indeed viable. Each one seems to be able to contribute one or more characteristics that could very well sum up a concept of the universe that might truly reflect observations, satisfy classical physics, and, ...be palatable. A realization is that each scenario offered is, and must be, speculative. They have to be, because all such visions are. The limitation of measurements and testing simply don't allow cosmologists much more than speculation. That there is more than one viable model attests to the lack of validation for any. But if speculation is a determinant of universal models, then you and I can speculate also.

The Meta model advocates deductive reasoning, so let us try that.

"Our" Universe:

Any model of the universe must indeed be palatable, because it has to be something that we can accept, something that we can use; it must be in keeping with our experience in this world and not offend our sense of logic and reason. It has to work. And, if we are going to all the trouble of designing a universe, then it becomes not just *the* universe, but *Our* Universe. We can begin by discussing those just mentioned "characteristics to discuss."

"That there was no beginning, and there will be no end, to the universe."

The QSSC model has the universe expanding for at least a trillion years, while the Steady State has it expanding forever. Because we have dismissed a big one-time explosive beginning, and because a quieter instant beginning is not any likelier, and because it certainly could not have gradually come to be, the universe *must* have been in existence for a long, long time, and most probably for an infinitely long time.

We truly can not conceive of any beginning at any time, because there is simply no mechanism for such a happening. Nature, and that grander Order, has simply never provided for such a thing; indeed, perhaps they couldn't. Order cannot be turned on or off as with a switch. And an infinitely long time is not at all foreign to the concept of relativity, because time, too, is relative and really not an issue.

It follows then, that:

a. There was no beginning.

The concept of an end is a different concern. Nature provides for dying as a natural pastime, and the laws of entropy are supposed to demand an end, ...some time. But, if we think of *nature* as an Earthly concept, or at least one that is of living things, then nature is not the dominant consideration in any universal model; Order is. That gander Order that defines each particle and wave, and controls every aspect of reality, from one end of the vastness to the very farthest reaches of the other; is really a set of laws that are without time constraints. Order is not the opposite of chaos in this discussion, because our grander Order provides for chaos as the stuff from which universal assemblage and eventually nature, are born. That grander Order, that one perfect law, then, does not have a shelf life.

Entropy is tough: demanding an irreversible, seemingly inexorable flow from order to disorder; a destined slowing, and cooling and decay of all things. But. Entropy doesn't seem to work very well, or perhaps at all, in living things. All life, even that life which might be questionable, seems to suffer an unrelenting compulsion for increased complexity, and that is not in keeping with the laws of entropy.

And there is plenty of that "life which might be questionable" in the universe. Guy Murchies's *The seven Mysteries of Life* tells us that there is no scientific differentiation between life and non, and shows us, beautifully, that even the simplest crystals exhibit attributes that we normally assign to an only living things. A universe, Our universe, that is chock-full of this life and questionable life then, need *not* fall victim to entropy.

> Is that a jellyfish or is it a colony of mutually supportive, highly organized, separate tiny animals? It that good looking gal a gorgeous creature, or is she merely a symbiosis of cellular disparities that in themselves are composed of chemical compounds not ordinarily thought of as alive?

Entropy may not have its way, after all.

It follows then, that:

b. There will be no end.

"The Universe is expanding."

The Steady State and QSSC models agree that Hubble's observation of, and Einstein's acknowledgment for, a pervasive expansion are correct; and that the expansion has been going on for a very long, perhaps an infinitely long time. The redshift indicators are almost unanimously accepted as proof, but as we have shown, not everyone agrees; and the Meta model at least, says that it is not.

We have shown that redshift may be merely perceived; that there is plenty of reason to question its reliability; that expansion might indeed be triggered by explosions of matter and antimatter or QSSC's mini-bangs; that expansion, even if real, needn't be related to a Big Bang.

Since the Big Bang is discredited, another explanation for the apparent expansion is required.

The presumed expansion of the locally observed universe, which is admitted to being a mere fraction of the total, does not mean for one moment that the entire universe is so effected. It seems reasonable that some parts of the universe could expand while others contract. If one part of space has a higher density or is warmer than its neighbor, then a temporary expansion would naturally be experienced. The periodic introduction of new material, such as virtual particles, would surely force space to expand, because Einstein told us that mass distorts space.

It follows then, that:

c. Expansions and contractions are undoubtedly regional phenomena.

"Microwave background radiation is a normal, perhaps local characteristic."

We have pointed out a number of possible explanations for the observed background radiation: microscopic metallic "needles" or small filaments of plasma reflecting and radiating ambient microwaves; cosmic dust particles, or even a local matter and antimatter explosion. No one knows. These explanations seem reasonably natural, and perhaps they all contribute to what is surely a natural, on-going characteristic of space.

It follows then, that:

d. The observed microwave background radiation is a natural phenomenon.

"The Background temperature is natural."

The Argument, hopefully the realization, that the background radiation is a natural phenomenon, asserts that its temperature must be natural also. That outer space would have a temperature just a few degrees above absolute zero is no surprise whatsoever. You would have guessed that and so would I. That the observed universal temperature would not be perfectly smooth, but would vary to a small extent, depending on density or regional activity, seems perfectly understandable.

It follows then, that:

e. The observed background temperature and its variations are natural phenomena.

"Creation is continuous."

This concept is offered to us by the QSSC model of the universe. Instead of a one-time all encompassing event, the universe is said to experience a series of "mini-bangs," – in localized "creation fields" around galaxy clusters – of different energies, continuously agitating and restructuring the universe. And these have been occurring at erratic intervals over all time, such that matter continually appears and disappears in such a way, that, as the universe expands, its density remains constant.

Consider:

A "series of mini-bangs," in lieu of that one big one, might at first seem like a cop-out. Not so. There is indeed a fundamental difference between that one big wormhole birth, from nothing, and an on-going constant belching of material from fluctuating energy fields.

"Around galaxies of different energies." The high-energy extravaganzas that are popularly attributed to black holes are all in a galaxy of some sort. Our mini-bang belching of material does seem related to galaxies, but it is worthy to ask; which came first? Was the mini-bang material introduced because there was a galaxy present, or did the galaxy itself constitute the material that was introduced by the mini-bang? QSSC argues that the presence of mass acts as a catalyst for the introduction of new material, but that cannot be. The wide-ranging forces that result in the precipitation of mass must be triggered by a "need" for mass: an absence of mass where needed to balance things. The presence of mass could not serve as a trigger for more, because that kind of stimulus would result in all mass being dumped in the same place. No, nature, and Order, abhor a vacuum; needs are satisfied, balances are maintained. The different energies observed, of course, are determined by the strength of the belch that provided the necessary relief.

As we implied earlier, the concept of more *space* being introduced to the universe everywhere, universally, simultaneously, pervasively, uniformly, homogeneously; would surely be impossible in an expanding universe, but might make more sense in a steadier system.

But more then that, a concept of more *matter* being continuously introduced universally, in different places, simultaneously; in an on-going never-ending satisfying of the venerated conservation of energy (to replace all that energy being expended) seems quite reasonable. If we can create one atom of hydrogen per cubic mile per year, we can sustain a continuing universe.[236]

"Continuously agitating and restructuring the universe. Of course! All of the drama, trauma, and constant interaction that our Earth experiences constantly are a natural experience everywhere. The birth and death of galaxies follows the same laws as do the birth and death of tadpoles.

"As the universe expands, its density remains constant." Well, if we just agreed that expansions and contractions are undoubtedly regional phenomena, then temporary regional deviations in density are acceptable also, and the consideration is moot.

It follows then, that:

f. Creation is continuous, as needed.

g. Constant agitation and reconstructing does indeed go on.

"Space is not curved."

Einstein said that it is. The mathematics is clear. A "curving," or lensing effect, has indeed been demonstrated such that almost no one questions this. Meta, though, does. The Meta model relies on classical physics for its development, and presents a viable argument for a more "Earthly" explanation: that the "light carrying medium" is denser near mass, and light is effected much like it is in the denser medium of water, satisfying the observations. Quantum physics, remember, says that the limit to the speed of light does not pertain to speed of the early expansion of space![225] This, though, is a tacit admission that space *is* a "light-carrying-medium." This discussion is not moot. The idea of a light-carrying medium is not all that far-fetched. It does seem that light should require *something* to carry it; like air and water carry sound. And! There is now evidence that light might become *polarized* as it travels through "empty" space, more so in some directions than others.[242] This is important evidence *for* a light-carrying medium. Remember, when we were discussing relativity, we pointed out that it might not apply in the deeper parts of empty space.

Yes, the concept of a light beam being curved was tested during an eclipse of our largest local gravitational source: our Sun; and, lo and behold, bending was observed. The measured amount of starlight curving or bending wasn't significant; it was detected only in the most careful measurements; in fact, the observations were hopelessly ambiguous, and "..it took many decades for scientists to agree on how much light really bends in a gravitational field,."[297] but the result is considered "proof" of the universe's curvature.

Well, ...in addition to Meta's denser medium, we might discuss the consideration that the bending could instead, be that quite ordinary electromagnetic *deflection* that we are all familiar with; the deflection of electric currents by electromagnetic sources: the same phenomenon that drives all of our electric motors! Light beams *are* electromagnetic "currents," after all, and our good old Sun is one stupendous electromagnetic source! If the abstract mathematical concepts of relativity and the esoteric characteristics of an unworldly space can be made more ...worldly, ...by classical arguments, ...then so should we lean. If we accept this concept, then empty space need not be mystical or mysterious, and Our universe can be more straightforward.

It follows then, that:

h. Space is not meaningfully curved, but more straightforward.

"Relativity, the speed of light, and quantum mechanics can be explained."

Meta is not alone in trying to put a more classical face on relativity. We all feel more comfortable with more straightforward explanations. Even the maximum speed of light seems understandably limited by a medium; much like the speed of sound is limited by its medium. Relativity does indeed show that measurements, to include time, are effected in strong gravitational fields, but the *why* is not that evident. Classical arguments might work yet.

Quantum mechanics, and its acknowledged uncertainties, might, as we have hinted, be merely temporarily frustrated by its disassociation from classical physics. It seems a worthy hope that in the long run, normalcy will prevail, and since this is what we want for Our universe, especially since particle theory may well be invalid for situations outside the atom, we can deem it so.

It follows then, that:

i. Relativity, the speed of light, and quantum mechanics can be explained by classical physics.

"Redshift is explained."

Meta argues that redshift is due to light's consuming travel though its just discussed light-carrying medium, which seems pretty reasonable. This, plus all of the earlier criticism of the redshift concept, reinforces the idea that we can ignore it.

It follows then, that:

j. Redshift is irrelevant to the issue.

"Gravity is explained."

Tom Van Flandern's *Dark Matter, Missing Planets, & New Comets,* and his *Meta Research Bulletin* posit that gravity is the instantaneous action of Classical Gravitons (CG's) on mass, that these particles travel much, much faster than the speed of light, and that gravity has a limited range: about one kilo parsec, or about 3,000 light-years. Considering that a popular figure for the diameter of the Milky Way is about 100,000 light-years, 3,000 isn't much. But, such an extremely short range for gravity is not as incredulous as it sounds, if we accept both the idea that gravity is not the determinant in the universe, and the realization that all forces are more likely "pushing" things than they are "pulling" things.

Most scientists completely ignore such thinking, and embrace the doctrine that gravity is merely the distortion of space by mass. Poirot is corroborated. Debilitating dogma is no help. An explainable, classical gravity is handier and more palpable, especially since contemporary gravity seems not to work well in their universe. A "pushing" gravity does make more sense than a "pulling" one, and its limited range is not offensive. Van Flandern's explanation will do until a better one comes along.

It follows then, that:

k. Gravity is explained.

"String theory is representative of, and cosmic rays and star and galaxy formation are explained by, electromagnetic forces."

We have shown that string theory might be more applicable to liquid crystals, super-fluids, and plasma than it is for the cosmos, and that this mathematics, too, might be misapplied. Their interpreted enormous energy and fierce vibrations are quite descriptive of plasma fields, and seem to validate the concept of electromagnetic forces as a dominant factor in space.

The incredible energy of cosmic rays, again, might be better explained by their being propelled by extraordinarily strong interstellar magnetic forces and their picking up of more and more momentum as they pass through all of those huge magnetic sources that are everywhere in the universe.

And, the formation of stars and galaxies seem to cry out for an electromagnetic explanation. The collapsing cloud thing just doesn't work very well for stars, and even less so for galaxies in a conventional gravitational universe. A number of rationales, now, show the clearly electromagnetic, dynamo characteristics of galaxies, supernovas, and even stars; and the seemingly weak contribution of gravity.

It follows then, that:

1. String theory is representative of, and cosmic rays and star and galaxy formation are better explained by, electromagnetic forces.

"The fifth dimension of scale is born."

Not really. When I was a kid we came to the conclusion, and perhaps you did too, that the solar system was probably just an atom in a larger system. The idea that we were all part of something much bigger was quite natural for us then, so Meta's dimension of scale was no surprise, ...but it *must be so*.

There cannot be an ultimate smallness nor an ultimate largeness. There cannot be such a thing as a fundamental particle, which, unavoidably, must be composed of still smaller sub-particles, and they in turn must be of even smaller sub-substances. And yes, even the observable universe, and the much deeper unobservable universe, is merely some of the stuff of a larger something else. Any argument otherwise is insupportable and preposterous.

"Does any grain or atom really need to be distinguished from a world? From a universe? Whose voice in this earthly node of flesh can declare with authority that our universe is not an atom of some unknowable larger megaverse outside it? And what atom anywhere has been prohibited from having a microverse inside it? Furthermore, is there any evidence that relativity does not pervade all dimensions, even transcending finitude so that ultimately space and time and self unravel into some sort of Infinite ...for ever?"[137]

It follows then, that:

m. There cannot be an ultimate smallness nor an ultimate largeness.

"Elementary particle abundance are naturally brought about after an infinite time."

Both the Meta and QSSC models recognize that elementary particle ratios and abundances are the result of an infinitely long, sustained evolution of matter and energy interactions. The on-going pulsating vibrations and the teeming, effusive flows of agitation that characterize a dynamic cosmos are justification a-plenty for the observed. Needless "predictions" based on guessed-at rates of star birth and deuterium manufacture; and estimates of early densities are of no interest in Our universe.

It follows then, that:

n. Elementary particle abundance is naturally brought about after an infinite time.

"Galaxy development is explained by Meta and Plasma models"

Meta does seem to explain how gravity could result in the spiral galactic arms that have puzzled scientists all this time, but it's the very concept of a limited range for gravity that permits such construction! It's a question of neighbor influencing neighbor instead of a broader, long-range central influence![125] But, regardless, ...electromagnetic forces do seem dominant.

What if the old disk stars and the old globular clusters, all about the same age, were formed (gravitationally) together, in the center-disk area, and what if the process that formed these old stars resulted in the spin that we see today? And what if that spin generated an electric field; which it should well do, since it is spinning in the inevitably charged gas fields that are spread throughout the universe by supernovae; and what if that electric field, rather naturally, formed a toroidal electromagnet? And what if that electric, toroidal-shaped field attracted some of that gas and dust and those iron filings that are spread everywhere by the supernovae, into those wispy, filament-looking, spiral arms? And isn't it in active, gas-and-dust-rich regions that new stars are formed? Wouldn't this all explain why we find young iron-rich stars in the spiral arms between the older, iron-poor stars and clusters? Wouldn't all this interaction between gravity and electromagnetism better explain our observations?

The above suggests that gravity could indeed, should indeed, support galaxy development, but only in a very limited way; and the concept of different forces working together in this process seems quite natural ...indeed.

But, what if that just mentioned spin-producing process was itself not gravitational but an electromagnetic discharge resulting from a belching, gushing conversion of energy to mass? QSSC credits galaxies with being the source for new material in the so-called min-bangs, and the plasma model seems to support that kind of thinking. Whether galaxies themselves trigger the precipitation of more matter from the energy fields, or they themselves are merely products of those energy fields, the concept of mass creation in and with galaxies seems quite workable.

And, the mathematics is there! The math that supposedly describes the creation of mass from nothing is essentially identical to that which would describe the condensation of matter from wide-ranging energy fields. It's a heck of a lot easier to precipitate protons from energy fields than it is to generate universes form nothing. The logic is overwhelming for those not afflicted by preconceived notions.

> "The idea of continuous matter creation ...does not seriously offend [Feynman] ...and that matter creation is possible because the rest energy of the matter is actually cancelled by its gravitational potential energy. 'It is exciting to think that it costs *nothing* to create a new particle,...' "[228]

It follows then, that:

o. Galaxy development is best explained as locations where matter is precipitated from energy fields into basically plasma configurations.

"Quasars are neophyte galaxies."

Those most mysterious and powerful, most brilliant objects in the universe, brighter than 10^{15} Suns; thought to be no bigger than our solar system, but emit more energy than 1,000 galaxies of 100 million, billion Suns each: very strong radio sources in addition to emitting infrared, x-rays, cosmic rays, and torrents of gamma-rays; .."these "improbable monsters" seem to lie at the heart of some galaxies."

NO!   They *are* the heart of some galaxies, and perhaps *were* the heart of all! Those same fluctuating energy fields that vary from the weaker mutterings of star creation, up to the thunderous ovation of galaxy production, are also responsible for the much larger, fresher torrential outpourings that we observe as quasars. These, then, also lend credence to "that vacuum, said to be thick with virtual particles."

Yes, the quasar is extraordinary to say the least. But only in degree! Quasars are merely the larger and NEWER gushes of energy from which all matter is born.[253] First would come the belch of material; not the lesser star-making "burp" type, but a mind-boggling torrent of energy that is popularly attributed to black holes. Next, but almost simultaneously, the forces of electromagnetism begin their shaping of the structure and the development of all kinds of elements, *even to the heavier*. Then, and only then, after the electromagnetic forces have weakened, the weaker gravitational forces are allowed to dress-up the structure into the more presentable, perky, usually spiral type of galaxy; such as our, *much younger than estimated,* Milky Way. Inevitably, though, after its tenured life span, it gradually decays into the shapeless cloud of residue called the "elliptical."

This must be so. Quasars are clearly sources of energy and material. They *are* the hemorrhages in the rich energy fields that pervade the cosmos from which all matter *must* come. This view of quasars and that of the electromagnetic development of galaxies espoused in Chapter 11, are clearly, mutually supporting.

Inadvertently giving credence to this idea, a team studying jets from quasars and active galaxies found "circularly polarized radiation from the jets [that] rotates around the wave's direction of travel." [288] You and I can cross quasars from that list of "Things That Are Admittedly Not Understood."

It follows then, that:

p. Quasars are neophyte galaxies.

And, …because galaxy development results from quasar inputs and an immediate start of electromagnetic structuring, the whole process can take far less time then estimated for the popular gravitational mechanisms. The fresh, sharp, "active" galaxy is so, only as long the quasars continue their output. Spiral galaxies, then, are much younger than thought, and our bright, perky; new looking Milky Way may be no older than the stars in it.

It follows then, that:

q. Galaxies are much younger than thought.

"Star birth parallels galaxy birth."

Not as obvious as it sounds. Of course, most stars are undoubtedly born in, and as part of, galaxies, with about the same naturally limited size as our Sun; ordinarily not much bigger than 10 times its size,[239] but *how* they are born is an issue. We showed that the "collapsing gas cloud" scenario was much too improbable, but that stars resulting from the collision of gas clouds was quite workable. The colliding-cloud mechanism, however, seems inadequate to explain all of the very many lone stars that wander through space, (A trillion or so in the Virgo cluster alone) [240] and another procedure might be called for.

Yes, colliding gas clouds must surely give birth to stars; individuals or clusters of them. Examples of this seem to have been observed. But those large belching gushes that convert energy into galaxies of different sizes and animation would also, surely, have their much smaller, gentler siblings in minor "burps," perhaps not quite sustained enough to form a galaxy, that could result in individual stars and groups of them. The orphaned stars then, might be subjected to being moved about by local streaming, ultimately finding themselves, as frequently observed, far afield from home. Those fluctuating energy fields must indeed fluctuate in intensity, such that the precipitation of mass from that fluctuation should naturally vary in degree of energy and its produce; just as a series of thunderstorms can vary from the brutal explosiveness of a frightful nearby strike, to the barely heard rumbles and barely seen ethereal flashes off in the distance.

It follows then, that:

r. Star birth parallels galaxy birth.

"Black Holes are irrelevant or do not exist."

If black holes, the explanation to every thing otherwise unexplainable, aren't necessarily the source of power for quasars, or any of those "unexplainably large" spectacles; then all of that mathematics, and all of that desperation cannot justify a need for such things. They make no sense in a grander scheme, so the heck with them.

It follows then, that:

s. Black holes are irrelevant or do not exist.

"Structures are merely the result of wave action."

Remember, one of the powerful arguments against the Big Bang was that the supposedly observed large-scale structures would have had to have taken far more time to form then their scenario could possibly allow. We showed, of course, that the unreliability of redshift measurements means that those structures might not even exist, but ...there probably are "configurations" ...of a sort, throughout the universe. Even if redshift measurements are correctly showing some type of formation, it is undoubtedly simple bunches of galaxies that have formed in an entirely different way than popularly envisioned. The issue is not how long it took to make these structures, but how they were formed!

The very nature of one large, naturally dynamic universe of on-going hectic activity suggests the concept of *wave* action, much like we experience here in every conceivable consideration. As the Meta model points out, the bunching that may be observed is almost certainly the result of the natural wave action in the fluctuating energy fields that permeate the universe and forms the sinusoidal pattern of density that we are surely seeing. *All* of those "observed" so-called structures that cross and weave through the far reaches of space for billions of light-years, are ebbing and flowing back and forth for all of time.

The mathematically supported concept of mass precipitating from fluctuating energy fields, and those very, very wide-ranging changes in energy fields, are undoubtedly changing in tune with universal-wide wave fronts that ever transverse the cosmos! "Structures" and their spacing then, are merely the density ridges and grooves of a universal hi-fi record. ...And, what a tune!

It follows then, that:

t. Structures are merely the result of wave action.

"There is one, and only one universe."

   The mathematics keeps spitting out an infinite number of universes, such as that one from which we are supposed to be a pinch of. Well, ...that might be O.K. because, if the scale dimension is correct, then, the universe has to be part of something bigger; and you can pick anything you wish for that bigger something: a super-large hot dog in a super-large bun, or a super-large pebble on super-large beach. Whatever. The point is that there are probably many, or very, very many of these super-large what have you's, and the multidimensional math can well be applied to those.

   I guess that you could say that is merely semantics. But remember our argument that if this universe was born form another, the Order and laws in each would have to be identical; they cannot be measurably different, so it makes more sense that they be part of one and the same. And, if we are part of a super-large pebble on a super-large beach, then our pebble must be quite like our neighbor and we both, along with every other super-large pebble on that super-large beach had to have been formed in accordance with the same grand Order. No test – not even a thought process – can lend credulity to any of this, but it beats watching television.

It follows then, that:

u. There is one, and only one universe.

   Well, then, doesn't it follow that we have just described "Our" universe?

   a. There was no beginning.
   b. There will be no end.
   c. Expansions and contractions are undoubtedly regional phenomena.
   d. The observed microwave background radiation is a natural phenomenon.
   e. The observed background temperature and its variations are natural phenomena.
   f. Creation is continuous, and as needed.

g. Constant agitation and restructuring does indeed go on.

h. Space is not meaningfully curved, but more straightforward.

i. The speed of light, and relativity and quantum mechanics effects on cosmology, can be explained by classical physics.

j. Redshift is irrelevant to the issue.

k. Gravity is of finite range and is meaningfully explained.

l. String theory is representative of, and galaxy formations are better explained by, electromagnetic forces.

m. There cannot be an ultimate smallness nor an ultimate largeness.

n. Elementary particle abundances are naturally brought about after an infinite time.

o. Galaxy development is best explained as locations where matter is precipitated from energy fields into basically plasma configurations.

p. Quasars are neophyte galaxies.

q. Galaxies are much younger than thought.

r. Star birth parallels galaxy birth.

s. Black holes are irrelevant or do not exist.

t. Structures are merely natural variations in density and the result of universe-wide wave action.

u. There is one, and only one universe.

Well, at last, …there we have it. We have attempted to make a case against the Big Bang, and have offered a conjectured but satisfying alternative: a physical, classical, straightforward universe in which we can feel at home. And this is not mere wishful thinking; the evidence available, the difficulties in making meaningful measurements and determinations in the cosmos, and the best reasoning; negate none of this and support all of this, as much, at least, as for any other such notions.

Perhaps we will never know for certain, but an infinitely old, infinitely large universe, that somehow precipitates matter as needed from hemorrhaging energy fields, makes sense. Resplendent, sprightly, fresh-looking, spiral galaxies, …such as ours; with all of that extraordinary violent activity at the center, …that we observe; are the most vivid indicators of the on-going birth of matter and the intense electromagnetic activity that pervades a dynamic cosmos. Gravity, the very weakest force in nature, can be only a bit player …on this grand stage …in this magnificent show.

Kudos to the director.

Chapter 22

Why?

Why?

Why would an entire discipline go off on to the wrong track?

How?

How could they be so completely fooled?

Cosmology's regression to metaphysics and philosophy may well have begun around 1900, with Hendrik Lorentz' electron theory, which inspired Hermann Minkowski's ideas of four-dimensional spacetime, which was probably the basis for Einstein's special relativity.[246] The other shoe fell around 1917 with V.M. Slipher's determination that other galaxies were receding rapidly from ours, which C. Wirts related to distance, which Edwin Hubble put into the quantitative framework of his famous law.[255]

So, ...the relationship of redshift to recession velocity was instituted, the expansion of the universe was ratified, and popular cosmology traced that expansion back to a singularity.[256] *There* was the mistake.

Cosmological observations are not at all definitive because their meanings have to be inferred and they are not subject to test, so the telescopes were soon augmented with computers, and to a degree replaced by computer models. Astronomers, from that time on, were mathematicians more than observers, and they were off and running, ...all the way back to that impossible ...singularity.

> "Mathematical concepts, be they simple arithmetic or infinite set theory, make sense in the abstract, yet when applied to real things, their logic frequently becomes quite problematic."[262]

Richard Feynman, the most renown of physics teachers, cautioned earlier that; "We must be careful to interpret the results of our theories when they are treated with full mathematical rigor [or] the full mathematical rigor may convert ...errors into ridiculous conclusions."[231] But, that venerated teacher has been ignored by his pupils. Mathematicians are not at all deterred by singularities, but scientist *must* be. Singularities are impossible in nature and merely imaginary in science. The pristine virginity of pure mathematics is addictive for the hatching of grander schemes, and the pragmatism of reality tends toward the burdensome, ...for the possessed. The arbitrary use of any and all available mathematical theories: Higgs fields, quantum theories, string theories, GUT's and TOE's; in spite of their possible irrelevance, is the larger failing of the cosmologists.

Einstein's reaction to Hubble's law of redshift connotations resulted in all that cosmological constant confusion, and possibly caused a temporary distraction from the traditional observational, experimental characteristics of good science. And, once the intoxicating nectar of mathematical and theoretical methods became ritualized, the user became addicted. Now, there was no escape. The practice promised hallucinogenic, illusionary answers to irrelevant questions, but the believers felt that they were becoming "all knowing." The believers then became the preachers, and heretics were turned out. From then on, the myopic, dogmatic, entrenched insistence on a grandiose but clearly erroneous paradigm was "...*the way.*" Sounds irreverent, doesn't it? But, if you read the many quotes in the text and appendix, no happier interpretation is likely.

Einstein's veneration of gravity led them to ignore all other, far more meaningful forces, as they began fabricating the Big Bang models, all of which needed correction, ...and continue to. Because observations unrelentingly belied theory, all of that exotica was contrived: dark matter, black holes, great attractors, and that whole Chapter 13 list of imaginings. Then there were the "special" modifications, such as those special "early" stars and our very own Milky Way's special black hole. Spacetime was assigned a mystical and magical quality, and treated as a celestial womb from which all universes are born. As it became obvious that even these inventions were inadequate, conceived forces, such as the cosmological constant, cosmic strings, "odd" density waves, and now, anti-gravity and "quintessence" were envisioned.

Redshift, of course, has long since been shown to be an unreliable indicator of velocity or distance; the microwave background radiation and crucial light-element ratios have fared no better; and each and every proclaimed measurement is simply unreliable.

The scientific method and its principles, worthy competitive funding, and a truer education of our students will suffer until the course of today's cosmology is reversed.

Chapter 23

Now, What?

* Outlandish Explanations * Computer Models * Ambitious
Projects Planned * Etceteras * Telescopes and Satellites *
Most Distant Bodies Discovered *

Outlandish Explanations:

"Hardly and astronomical announcement makes the front pages without
being said to overturn all existing theories." [283]

Probably unnoticed by the average reader, that first phrase in a recent
article on gamma-ray busters is actually profound. The article relates to
gamma-ray busters, which are short duration, concentrated bursts of
gamma-rays that are observed about once a day, in all parts of the sky:
a true phenomenon, a true mystery. For one such event, "To be so
bright at such a distance, the burst must have shown more brilliantly
than any object previously recorded." Indeed, NASA called the event
"The most powerful explosion since the creation of the universe."[283]
"The most intense burst of cosmic gamma-rays ever seen" was detected
in 1979 and has puzzled theorists ever since. The explanation? A new
kind of star: *the magnetar.* "Unimaginably dense, this star would have
a sold crust [of iron] covering an exotic liquid core" and have huge
magnetic fields that would heat up the surface causing it to crack and
allow the burst of gamma-rays.[284]

[The magnetar is a newly envisioned form of neutron star. The
prescribed implosion process of forming neutron stars, of course,
precludes any material remaining in the form of recognizable

atoms, because everything is supposed to be compressed into neutrons; but such limiting considerations are routinely dismissed when another answer is needed.]

That above first phrase actually relates to almost everything newly noticed in cosmology. Just about everything observed, almost daily, seems to require extraordinary, outlandish characteristics. While other disciplines periodically, quietly, discover reinforcing, supportive or corrective contributions to the prevailing thought-process, cosmology's newer "discoveries" are invariably mind-boggling, phenomenal and bizarre; such as: unseen, undetected "attractors" that "must" be the equivalent of 50 billion, billion times the mass of the Sun; or bigger; cosmic strings of "a thousand trillion tons of mass for every inch of length;" or quasars brighter than 10^{15} Suns. Black holes, dark matter, cosmic strings, false vacuums, anti-gravity, or a new form of energy that varies with time and space. **And it simply does not stop.** Such chimerical claims do make the front pages, but almost all scientists, in all other scientific disciplines, are characteristically skeptical about incredible explanations for anything.

Computer Models:

We talked about the limitation of universe computer models; that great reliance on assumptions must be made; that pre-conceived notions and bias are necessary; that far more data and detail than can possibly be gathered is required; and that a computer with far more capability and speed than has been demonstrated so far, is required; that shortcuts are essential. Well... After a year of preparation and months of running time, and by running in parallel all 512 work stations of one of the world's most powerful supercomputers at the Garching Computer Center in Germany, "For the first time, cosmologists have harnessed enough computing power to model the entire observable universe.[281] Beginning one billion years after the Big Bang, when the cosmos was almost perfectly smooth and uniform..."

Stop ....right ....there.

[What about the "problems that all relate to the very first fractions of that very first second?" "The setting of physical constants and "the making of protons, electrons, quarks and their kin" in that first instant? (Chapter 3) The vital prerequisites that at "About one year later, the broiling gasses would have cooled to a level experienced in star interiors?" That "at 300,00 years

after the event, light began to escape from that super-density?"
"That galaxies started making at 200 million years?" (Chapter
7) These "events" are not mere chitchat; they are imperatives in
the scenario development. To begin a simulation well *after* the
most crucial milestones in the fabrication is a tacit admission of
weakness. ...When *was* Order begun?]

"The new simulations follow a *type* of slow moving dark matter,
cold dark matter..."[281] [Of course, it you introduce a cold dark matter
input, you will get a cold dark matter output. We all realize that
computer products are no validation of computer inputs, but it is easy to
forget.]

"In a second set of simulations, cold dark matter takes a back seat to
the energy associated with the cosmological constant..."[281] [Again, the
constant is treated as a force of energy, not as a mathematical symbol,
and cosmologists, of course don't know what the value of that constant
is.]

"To directly incorporate galaxy formation, the models would have
to include effects other than gravity, notably gas pressure, heat and
radiation."[281] [But, ...they didn't! ...Nor electromagnetism! ...Nor
the possibility that the Big Bang may be wrong! It is easy to forget that
computers are mere tools: tools that can be misused.]

Ambitious Projects Planned:

Sloan Digital Sky Survey: "The most ambitious mapping of the
heavens ever undertaken, is about to start from [Apache Point, NM]
using the most complex camera ever built." Taking pictures in three
dimensions, [relying on redshift for distance, of course] the $77 million
project will collect an amount of data equal to what is now stored in the
Library of Congress.[278,280]

[Does this mean that we won't have to build anymore telescopes?
What will we do with all that data when redshift is admitted to be
unreliable?]

NASA has just awarded a $600,000 contract to duplicate the
experiments of a Russian scientist who claims to have invented a device
that blocks the force of gravity. NASA is paying an Ohio based
company...to continue a series of experiments on gravity shielding.[295]

[Fill in *your* comments]

Etceteras:

"New measurements ...provide "unambiguous" evidence that the Milky Way's core contains a black hole as massive as 2.6 million Suns." Because dust obscures visible light, infrared observations were made."[260]

"A Dwarf galaxy that has been orbiting, and "diving right through" the Milky Way without disturbance, for 10 billion years, has been discovered. Something else is holding it together, otherwise the little galaxy wouldn't be there any more. The 'glue' is probably a hefty helping of mysterious dark matter."[265]

"Scientists studying exploding stars more than 7 billion light-years away have found evidence of a *mysterious antigravity force* that is causing the universe to expand at an accelerating rate."[269]

"Very old and distant galaxies are still being detected."
"We are seeing galaxies in the very early universe as strongly clustered as they are today."
"This is the beginning of a huge new effort, which is going to occur worldwide."[257]

If the curved, finite universe is as mathematically described, what astronomers think is a distant galaxy *could actually be the Milky Way seen* at a much *younger age!* To verify this, it is proposed that a search for "circles" of matching temperature variations be undertaken. The launch in a few years of the sun-orbiting Microwave Anisotropy Probe promises to make a quest for circles feasible.[271]

The apparent lack of sufficient density to support a flat universe, in spite of all of the conjectured dark matter, is prompting a search for "other" forms of matter or energy, which could make up the difference. It has prompted a suggestion that the cosmological constant might decrease with time and vary with location; and, ...a new form of energy called *quintessence*, which adds a cosmic energy density that varies with time and space is imagined.[274]

A recent COBE study of the far infrared, after estimating and subtracting the considerable infrared signals from the solar system and Milky Way's dust and heat, resulted in the determination that dust absorbs about two-thirds of the visible light emitted by galaxies, and ...it is believed that the infrared background comes from dust associated with distant objects making stars at a feverish rate.[272]

[Should the popular star-making process produce dust?]

Stars within the Milky Way's disk are much younger than those in its halo, so, it is determined that the two groups could have assembled separately. It is suggested that the halo stars could have been snared from neighboring galaxies.[273]

[How a younger bunch could snare an older, bunch, and from such vast distances, was not discussed. Completely disregarded is the admission that they simply do not understand galaxy development.]

Most Distant Bodies Discovered:

Periodically, enthusiastically, new, "most distant" bodies are reported, with considerable fanfare.

Nov	'89 [15]	14 billion light years
May	'91 [30]	12 billion light years
June	'91 [77]	10 to 12 billion light years
May	'94 [147]	12 billion light years
Oct	'98 [289]	12 billion light years

Telescopes: And Satellites:

Almost every new "discovery" is a rephrasing of the others. Additional so-called evidence, of more of the same inferred possibilities, demands further study, ...and more telescopes. "This is the beginning of a huge new effort which is going to occur world wide."

"Testing this notion will require a new generation of telescopes, one that can resolve individual galaxies in the smooth infrared background. A slew of instruments...scheduled for launch early in the *next* decade may provide answers."[264]

"As the millennium approaches, it is NASA, Chief Dan Goldin's goal to '*blacken the sky*' with spacecraft."[261]

"So many are planned, that by the year 2000 Palomar will only be the 15th largest." [Telescopes on the moon and even Pluto are dreamed of.]

"The problem now is that there is going to be more telescopes available than money to operate them."[169]

They really are engineering wonders, so technically marvelous, so challenging to design and construct, that to work on their construction, and experience the pride of accomplishment when they finally come on line or on their way, is to be convinced of their value and the importance of their mission. You can't help being impressed. They are phenomena of form and function. But, these things cost a lot of money! Hubble cost an estimated 7 billion dollars. The Sloan Digital Survey will cost over 77 billion, in spite of the realization that its key parameter, redshift, may be completely misunderstood.

We are willing to fund research that is meaningful, professionally done, and intended for our betterment; but, if observers are myopic and merely reinforcing preconceived notions, it is just possible that *we don't understand what we see*! If observers do not employ the scientific method, if they are unwilling to entertain questions, it is just possible that *we may never understand what we see*!

There are hopeful signs that the list of imaginings in Chapter 13 is simply too long; and now, quintessence and anti-gravity are simply too much, for those scientists that are beginning to realize the impossibilities of the current popular scenario. Hopefully, those wanting to return to the scientific method might yet get a chance on those telescopes. Perhaps, then, Webster might define cosmology as a science.

In Final Perspective:

If the dimension of scale is viable,
And it seems inescapably so,
Then we,
Like Gods of this scale when we reflect with
Our wider view on the smaller-scale universes
In a rather ordinary small stone, might better visualize our own.

The ordinary can seem mystical in that such
Wonders can seem ordinary!
And we,
At least, might share with our micro-kin
A mystical wonder of mystical,
Ordinary Order.

Appendix I

Selected Quotes

"When Big Bang proponents make assertions such as 'an expanding universe...very well verified observationally,' 'a whole bunch of observations that hang together' and 'the evidence taken together hangs together beautifully,' they overlook observational facts that have been piling up for 25 years and that have now become overwhelming. Of course, if one ignores contradictory observations, one can claim to have an 'elegant' or 'robust' theory. But it isn't science.

"One point at which our magicians attempt their slight-of-hand is when they slide quickly from the Hubble, redshift-distance relation to redshift velocity of expansion. There are now five or six whole classes of objects that violate this absolutely basic assumption. It really gives away the game to realize how observations of these crucial objects have been banned from the telescope and how their discussion has met with desperate attempts at suppression."

"The alternative to the Big Bang is not, in my opinion, the steady state; it is instead the more general theory of continuous creation. Continuous creation can occur in bursts and episodes. These mini-bangs can produce all the wonderful element building that Fred Hoyle discovered and contributed to cosmology. This kind of element and galaxy formation can take place within an unbounded, non-expanding universe. It will also satisfy precisely the Friedmann solutions of general relativity. It can account very well for all the facts the Big Bang explains – and also for those devastating, contradictory observations which the Big Bang must, at all costs, pretend are not there."

Halton Arp, Astrophysicist
Max Planck Institute for Astrophysics
Munich, Germany
p.51,27 July, 1991, *Science News*

"Ivars Peterson's article illustrates once again that explanations of the steady-state model are clear and demonstrable, while those of the Big Bang are incoherent and shrouded in mystery."

Scott Nicholson
Bradenton, Fla.
p.51,27 July, 1991, *Science News*

"Cosmologist David Schramm has his analogies a bit mixed up when he says that the new discoveries of huge super-clusters of galaxies cause no more problems for the Big Bang than our inability to predict tornadoes causes for the idea that the Earth is round. The problem with the large-scale structure found in the universe is that it is clearly 10 times *older* than the 10 or 20 billion years the Big Bang allows for the age of the whole universe."

"The vast ribbons of galaxies recently observed are separated by nearly a billion light-years of space. So to form them, matter has to travel half that distance, if the universe was originally smooth. But galaxies are observed to travel at only $1/600^{th}$ the speed of light, so the huge structures must have taken at least 200 billion years to form. Having 200-billion-year-old structures in a 20-billion-year-old universe causes the same sort of problems as having millions-of-years-old mountains in a 6,000-year-old Earth."

Eric J. Lerner, President,
Lawrenceville Plasma Physics
Lawrenceville, NJ
P51,27 July, 1991, *Science News*

"I learned in graduate school that due to the equi-partition of energy, the energy densities in the galaxy of magnetic fields, gravitational fields, and kinetic energy are comparable in magnitude. The large energies associated with magnetic fields (plasmas) must have some effec*t* in the structure of galaxies."

"I can understand why computer simulations of galaxies ignore plasma effects, due to the difficulties of incorporating such effects in the programs. But I cannot understand why theoretical discussions simply ignore plasmas."

"The most pleasing theory of cosmology from an intellectual standpoint is the steady-state theory of Fred Hoyle. It is based upon the strong symmetry principle – namely, that there is symmetry in the large in the universe in spacetime. There can be no fundamental difference in the universe here or anywhere else, or at any time. This theory lost favor due to difficulties with observations. However, recent observations and modern theoretical developments may yet rescue it."

"There are other theories of physics that negate the Big Bang. The gravitational theory of Nathan Rosen, for example, denies the existence of black holes and the Big Bang."

"The interesting question is: How do certain theories become so accepted that other theories of equal validity cannot even be heard? How many papers dealing with such other theories are rejected by referees? Is the Big Bang in favor due to religious feelings?"

Sanford Aranoff
Kiryat Motzkin, Israel
p.51,28 July, 1990, *Science News*

"In a concluding salvo pitched at the Big Bang, Arp and his coterie of elderly radicals write, 'As a general scientific principle, it is undesirable to depend crucially on what is unobservable to explain what is observable, as happens frequently in Big Bang cosmology.'

'I [Eric J. Learner, an independent researcher in Lawrenceville, NJ] found that, indeed, as you go further away from Earth, the amount of radio radiation for a given amount of infrared radiation falls,' Lerner says. 'Unless you can come up with a very good explanation of why radio galaxies further away from the Milky Way are such wimps, then you have to admit that there is a lot of absorption going on. And if you can't explain the data, then you simply can't use the microwave background as evidence for the Big Bang.' "I view the Big Bang itself as a woefully incomplete theory." [Paul J.] Steinhardt [University of Pennsylvania in Philadelphia] says.

Ivars Peterson
"State of the Universe"
p.232, vol. 139, *Science News*

Langmuir's suggested six criteria of pathological science:

(1) The maximum effect that is observed is produced by a causative agent of barely detectable intensity, and the magnitude of the effect is substantially independent of the cause.
(2) The effect is of a magnitude that remains close to the limit of detectability or, many measurements are necessary because of the very low statistical significance of the results.
(3) There are claims of great accuracy.
(4) Fantastic theories contrary to experience are suggested.
(5) Criticisms are met by *ad hoc* excuses thought upon the spur of the monument.
(6) The ratio of supporters to critics rises to somewhere near 50% and then falls gradually to oblivion.

John R. Huizenga,
In *Cold Fusion; the scientific Fiasco of the Century*,
quoting Irving Langmuir, *Physics Today*, October 1989.

"Burbidge cites the case of Halton C. Arp, an astronomer who was denied telescope time at Mount Wilson and Palomar observatories in California because his observing program had found and continued to find evidence contrary to standard cosmology. Such sanctions against dissidents from the party line are not unusual, according to Burbidge. Unorthodox papers 'often are denied publication for years or are blocked by referees.' Even worse, he says, the same censorious attitude applies to academic positions."

"The Big Bang Censorship"
p.22, 13 April, 1992, *Insight*
An article by Arnold Beichmann
quoting the February 1992 issue *of Scientific American*

"Those who welcome the idea of a "beginning" forget that all one can assuredly say is that this is a state of high density of matter quite distinct from the distribution of isolated stars known to us; one may doubt that in this state the notions of space and time are applicable, because these notions are intimately related to the dispersed system of stars. The 'beginning' refers only to our ability to describe the state of things in terms of accustomed concepts. Whether there was creation from nothing is not a scientific question, but a matter of belief and beyond experience, as the old philosophers and theologians like Thomas Aquinas knew."

> From Max Born's *Einstein's Theory of Relativity*, revised.
> (Dover Publications, New York, 1962), p.369.
> Given to the author by Emil Wolf, Wilson Professor of Optical Physics,
> The University of Rochester.

"Does any grain or atom really need to be distinguished from a world? From a universe? Whose voice in this earthly node of flesh can declare with authority that our universe is not an atom of some unknowable larger megaverse outside it? And what atom anywhere has been prohibited from having a microverse inside it? Furthermore, is there any evidence that relativity does not pervade all dimensions, even transcending finitude so that ultimately space and time and self unravel into some sort of Infinitude ...for ever?"

> Excerpted from *The Seven Mysteries of Life*
> Copyright © by Guy Murchie.
> Reprinted by permission of Houghton Mifflin Co.

"Geoffrey Burbidge, professor of physics at the University of California at San Diego...has charged [in an astounding attack on the internal politics of astronomical theory] that 'conformity' is the order of the day in the field of cosmology..."

"...It is extraordinarily difficult to get financial support or viewing time on a telescope...unless one writes a proposal that follows the party line."

..."[The Big Bang] rests on, however, many untested, and in some cases, intestable, assumptions." The result is a "bandwagon of thought that reflects faith as much as objective truth."

"...there are good reasons to think the Big Bang model is seriously flawed."

"A remarkable set of observations by Tyson et al. show that the distribution of the dark matter in tested galactic clusters closely parallels that of cluster red light (baryonic matter). The most straightforward interpretation of this result is that the dark matter is baryonic. Given that there is no empirical evidence for non-baryonic dark matter, and some evidence against it, should it still be our best candidate?"

"A curious dynamic tension had been rising in the field of cosmology. Some widely held theoretical assumptions are coming into increasing conflict with observational results, and yet those assumptions continue to receive strong support..."

"...surely science would be better served if a greater attempt was made to loosen the grip of prevailing prejudices."

Robert L. Oldershaw
Amherst College
Amherst, Massachusetts
p.800, vol. 346, 30 August 1990, *Nature*

"The observations suggest that younger matter is intrinsically redshifted. The first consequence of this is that extragalactic redshifts are not velocities, the universe is not expanding, and the big bang did not create everything out of nothing. The only theory that can explain all these empirical results so far is the variable mass theory of Hoyle and Narlikar. This requires the intrinsic redshift to be a direct consequence of the episodic creation of matter in an initially zero-mass state."

The first paragraph of Summary, Halton Arp's essay, 1995,
Observational Cosmology Impacts Physics:
Given to the author by the great man himself.

Appendix II

Definitions Handout

Absolute zero: No molecular motions. 0° Kelvin, -273° C.

Antimatter: The reverse of conventional or baryonic matter. The proton is negative and the electron (positron) is positive.

Baryonic: Conventional matter. A class whose most important members are the proton and neutron. The total number of baryons in a system is said to be conserved, which means that the proton must be absolutely stable.

Black hole: Proposed collapsed star so dense that not even light can escape its powerful gravitational field.

Brown dwarfs: Thought to be too small to attain nuclear burning. Since they don't burn like stars, they are very dim and very difficult to find. One or two candidates seem to have been spotted.

Cepheids: Bright, slowly pulsating stars whose brightness is thought to be proportional to their pulse-rate. Scientists, then, use the pulse-rate to determine their intrinsic brightness, and use them as cosmic yardsticks.

Classical gravitons: Proposed by Meta Universe proponents, see Chapter 20: Very much faster than the speed of light; they push masses together rather than pull, such as envisioned in popular gravitational theories.

COBE: Cosmic Background Explorer Satellite. Launched November 1989. Measured variation in microwave background radiation.

Cosmological constant: Invented by Einstein to counteract gravity and keep the universe from expanding. Proportional to the vacuum energy density. Relied upon heavily as a conceived force to explain the energy density and microwave background radiation.

Dark matter: Hot and cold. Proposed invisible particles that would provide the additional gravity needed to keep galaxies from flying apart and in general support the Big Bang concept. Cold dark matter is supposed to move slowly, but hot dark matter is said to move at nearly the speed of light.

Domain: The volume inside a surface-like defect called a domain wall that supposedly forms between regions of different densities and velocities of matter, or between universes.

Entropy: A decrease of energy, or an increase in disorder: increases with time in the natural process, and the reverse is considered impossible: entropy decreases do not occur in nature.

False vacuum: A theorized (called bizarre) state of matter with a super high energy density. Something to make universes from.

Fermion: Particle that makes up matter. Divided into two subclasses: Quarks and Leptons.

Flat Universe: The energy density equals the critical density so the universe will ultimately stop expanding. Space is described by familiar Euclidean geometry: the way it looks to you and I. By far, the preferred version, but the odds in a Big Bang seem against it.

Gamma-ray bursters: Mysterious, unexplained, may be the most energetic phenomena in the Universe. Their symmetric distribution in the sky means that they must be very far away, which means that their intrinsic energy must be astronomical. The bursts last only seconds or minutes; a single burst emitting more energy than one quintillion Suns.

Globular clusters: Consist of 100,000 or more stars stuffed in a space just 100 light-years across. Many galaxies, though, including the Milky Way, are peppered with the things. Invariably, however, the stars in the clusters shine with the low-energy reddish light of extreme age. They are said to be the oldest stars in the Universe, as much as 16 billion years old.

Gluon: Carries the strong nuclear force that couples quarks.

Graviton: Hypothetical carriers of the force of gravity. Suggested to be able to drift off on their own in outer space to collect in galaxies. Might have to travel much faster than light.

Gravity: Said to be a matter of geometry – a consequence of the curvature of four-dimensional space-time.

Gravity waves: Einstein predicted that mass disturbances result in them. Not detected.

Great attractors: Supposed gravitational mass 50,000 quadrillion times the mass of the Sun that seems to be attracting great groups of galaxies.

Great wall: 1,700 galaxies, 500 million light-years long, a structure detected by redshift measurements. Unexplainable in Big Bang terms. It may not exist. If it does, theory can't explain it. If it doesn't, redshift may be in error.

GUT: Grand Unified Theory for Unifying the strong and electroweak forces.

Higgs field: An extraordinarily complex three-dimensional mathematical description of the Big Bang moment. Seemingly contrived.

Horizon: The distance light could have traveled by any one time. A problem is that the universe seems to have exceeded it.

Hubble constant: A number used to calculate the distance to stars based on their redshift. Related to the expansion rate.

Ia supernova: Ia supernovae are said to be the brightest in the universe; are said to have (unconfirmed) about the same intrinsic brightness; and are used in relative distance estimations.

Intensity values: Astronomers measure celestial brightness on a scale running from negative numbers, which represent the brightest objects; to positive numbers, which represents the dimmest. For example: The Sun measures –26.8; Sirius, the next brightest star in the sky, is –1.4. The faintest stars that can be seen without magnification measure +6.

Isotropic: The same, viewed form any point in space.

Jets: There are many jet-like flows of matter and energy seen emanating from stars and galaxies that are not well explained in gravitational terms.

Kech: Largest telescope in the world. Located in Hawaii, it has 36 hexagonal segments, each aimed individually by computer. Has a twin on Mauna Kea.

Kelvin: Kelvin degrees are the same as Centigrade degrees, but where the Centigrade scale assigns 0° to the freezing point of water, the Kelvin scale assigns 0° to absolute zero, where there is no molecular motion whatsoever. It's the equivalent of -273° centigrade.

Lensing: Einstein deduced that the gravity of a nearer object would bend the light from a more distant object.

Lepton: A basic particle: if charged, an electron, muon, or tau; if uncharged, a neutrino.

Light-year: The distance light travels in one year, 5.88 trillion miles.

Machos: (massive Compact Halo Objects) Proposed invisible small stars. Believed to be scattered throughout the galaxies in abundance, but still only makes up about 20% of the mass needed.

Magellanics: The local group of galaxies that are our most conspicuous companions.

MAX: A balloon-borne microwave experiment, more accurate (0.5 degrees) than COBE.

Max Planck Institute: For Extraterrestrial Physics in Garching, Germany. Where Magnetars (Chapter 13) were conceived.

Monopoles: Theorized isolated north or south magnetic poles; supposedly 10^{15} times as heavy as a proton. They probably do not actually exist, because they haven't yet been detected, and ...because my old physics book says that they don't.

Nucleon: The collective name for neutrons and protons.

Neutrino: Now thought to have a tiny fraction of an electron's mass,[279] can transverse the Earth without interacting with anything. Travels at near the speed of light.

Neutron stars: Tiny remnants of supernova detonations, which are so powerful, that protons and electrons, are condensed into extraordinarily dense cores of neutrons.

Nova: A star in the act of exploding..

Olber's paradox: Heinrich Obler argued that if stars are evenly distributed, we should see light everywhere we look. That we don't is known as Obler's paradox.

Omega: The ratio of the true average mass density of the universe to the critical density needed to just barely close the cosmos into a flat universe, which is ONE. The trouble is, in spite of cosmologist's desires that Omega equal One, the evidence does not support any value close to ONE.

Palomar: 200 inch Hale telescope on Mt. Palomar, near San Diego, CA.

Parsec: About 3 light-years.

Particles, fundamental: Two broad classes: Fermions and Bosons.

Photon: Quantum unit of light, caries the electromagnetic force.

Plasma: Ionized gas that can conduct electricity.

Positron: Positive electron.

Pulsar: Super dense, rapidly spinning, neutron star. Envisioned to emit a beam (or two) of radio waves as it spins at up to several thousand times a second.

Quantum mechanics: Another Name for particle physics.

Quark: The constituent of neutrons, protons, and related particles called hadrons. A "basic" particle.

Quasars: Short for quasi-stellar objects. Said to be what we see when a black hole swallows gas clouds and stars. Said to be only the size of our solar system, but they are the brightest objects in the universe, emitting more light than 2000 galaxies of 200 billion stars each. Unrecognized as energy and matter sources.

Quintessence: A new form of energy that allows the cosmic density to vary with time and place.

Redshift: Redshift is like the doppler-effect: the faster a galaxy moves away, the more its light 'shifts' to the red end of the spectrum – a lower frequency. Supposedly, you can tell how fast a body is moving away, and therefore how far away it is by the amount of its redshift.

ROSAT: X-ray satellite named for Roentgen.

Spacetime: Space, as time passes.

Speed of light: 186,284 miles per second, 300 million meters per second.

Spectrometer: Spreads starlight into its different wavelengths.

Star-burst galaxies: See Gamma-ray bursters.

Supernova: The explosive end of red-giant stars. Supposedly the source of most of the heavier elements.

Super-symmetry: Presumes that all the various kind of forces and particles in the universe are the same if you melt down their differences at a high enough temperature; the temperature of the Big Bang. All particles would come in pairs.

Symmetry: A demand of quantum mechanics.

Tau: Charged lepton.

TOE: Theory of everything. It would describe all the interactions of all the fundamental particles. Would join gravity with the GUT.

Virgo cluster: Centered about 50 million light-years away, encompasses 2500 galaxies, crucially used as a celestial yardstick.

Virtual particles: Pairs of a particle and its corresponding antiparticle that appear spontaneously, interact, and disappear. Contributors to the total energy density of the vacuum.

VLBA: Very long Baseline Array, Albuquerque. World's biggest. 5,000 miles wide, it stretches form Virgin Islands to Hawaii, up to 1000 times more resolution than any other, including VLA.

Weak force: A class of forces that are about 10^{14} times weaker than the strong forces that bind the nucleus. The neutrino is an example.

White dwarfs: Dense remnants of stars, most are no bigger than the Earth, but have up to a few times the mass of the Sun. Ultra violet emitters.

WIMPS: Proposed weakly interacting massive particles.

Wormholes: A bizarre small pinch of spacetime, tinier than particle, the mouth of a cosmic string. The name is usually interchanged with cosmic string. One mouth might or might not attach to a point in our universe; and the other mouth might or might not attach to a point in another universe. Might exist at the core of a black hole.

Appendix III

An Aside

Until recently, government money, your money, was readily available to those that could make a case for the Search for Extra Terrestrial Information; SETI, NASA, still, is planning a system of antennas to look for extraterrestrial life. Planned are antennas in orbit big enough to spot planets in other solar systems, and, "eventually, telescopes will be set up on the moon" ...and Mars ...and even one on Pluto is dreamed of. Scientists have been looking for extraterrestrials since 1961. Recently, Congress has cut off all funds for such efforts, but private donations are keeping the effort alive. Even now, in a number of observatories, teams of people sit around watching for and waiting for signs from intelligent life in outer space. ...And they watch, ...and they wait.

* * *

One day in the dark pre-pilgrim period, two angry stalwarts dragged one less so to the tent of the Great Chief of the Penobscots, on the banks of the large river that flows into the great sea.

"Who have we here? Said Chief Great Stuff of the Penobscots to his lieutenant, Big Red warden.

"Oh wise Chief, he is Early-sky-looker, of the Sagan lodge. Old Eagle Eye and I, bring this miserable creature here for your judgment on charges of treason."

"Treason?" said the great one, looking at the miserable creature kneeling between Big Red Warden and Old Eagle Eye.

"Mercy," pleaded the miserable lesser one, "All I did was throw a few clay bottles into the river."

The great chief looked quizzically at Big Red Warden.

"Oh Wise One, he was putting messages in those bottles that betrayed our location and welcomed visits."

"Why is that so bad?" said the grand man, "every one knows where we are, and this is tourist country after all."

Big Red Warden said, patiently, "But the bottles would flow to the great sea and across it to where it is rumored that aliens reside."

"So?" said the Grand Great One.

"Oh Great One, if there are aliens across the great sea, there can only be two types: those that are behind us in development and those that are ahead of us in development. Those that are behind us can't get here as tourists, so the bottles are merely wasted. Those that are ahead of us will come here; and with their superior technology take our lands, and introduce our young braves to firewater, and put us on reservations."

"Right. ..Right you are," said Great Stuff, the great, grand, wise Chief of the Penobscots. "Feed him to the dogs."

"Will do, oh Wise One,' said Big Red Warden, "may your wisdom forever protect us."

                    * * *

Ignored are the concerns of Big Red Warden and the impracticality of it all. Suppose we get a message from Alfa Centari, four light-years away. If you ask a question, it will take eight years to get an answer, and we'll both be under a new administration.

Appendix IV

Bibliography

1. "Satellite finds evidence of dark matter in galaxies" *Bangor Daily News*.
2. "Star Watch" Edwardo Citrinblum, *Discover:* 20 April 1993, p.20.
3. "Weighing the darkness" Sharon Begley, *Newsweek:* 18 January 1993, p.54.
4. "The Young and the Globular" Jon Holtzman/NASA, *Discover:* June 1992, p.24.
5. "Breakthroughs" Hume Feldman, *Discover:* June 1992, p.10.
6. "The Big Bang Censorship" Arnold Beichman, *Insight:* 13 April 1992, p.22.
7. "The Handwriting of God" Sharon Begley with Daniel Glick, *Newsweek:* 4 May 1992, p.76.
8. "Echoes of the Big Bang" Michael D. Lemonick, *Time:* 4 May 1992, p.62.
9. "COBE Causes Big Bang in Cosmology" M. Stroh, *Science News:* May 2 1992, p. 292.
10. "Gravity lenses for peering into darkness" *Science News*: May 2 1992.
11. "Big Bang and cosmology" Clair Wood, *Bangor Daily News*.
12. "Lumpy local universe unveils cold message" I. Peterson, *Science News*.
13. "The Creation of Big Bang Bickering" Richard Lipkin, *Insight:* 15 October 1990, p. 56.
14. "Cosmic Evidence of a Smooth Beginning" I. Peterson, *Science News*: vol. 137, p. 36.
15. "Astronomers find quasar that is oldest, most distant object in universe" *Bangor Daily News:* 20 November 1989.
16. "Cosmic Cartographers Find Great Wall" A. McKenzie, *Science News*: vol. 136, p. 340.
   "Quasar illuminates the most distant past" I. Peterson, Ibid.

17. "Seeding the Universe" Ivars Peterson, *Science News:* March 24, 1990, p.184.
18. "Enigmas of the sky: Partners or strangers?" R. Cowen, *Science News:* March 24, 1990, p. 181.
19. "Great Bubbles in the Cosmos" Michael D. Lomonick, *Time:* 27 November 1989, p. 57.
20. "Mix-and-Match Computing" Ivars Peterson, *Science News*: May 1, 1993, p. 280.
21. "Can We Really Understand Matter?" Eugene Linden, *Time:* 16 April 1990, p. 57.
22. "Galaxy map smoothes out the vast cosmos." R. Cowen, *Science News*: vol. 137, p. 262.
23. "Cosmological inflation: A budding universe." I. Peterson, *Science News*: vol. 13, p. 358.
24. "Quasar light points to younger galaxies." I. Peterson, *Science News*: June 23 1990, p. 389.
25. "Plasmas ignored." Sanford Aranoff, Kiryat Motzkin, Israel: "Letters." *Science News*: July 28 1990, p. 51.
26. "Too Smooth a Universe." *Discover:* January 1991, p. 34.
27. "Return of the cosmological constant." *Science News*: vol. 139, p.12.
28. "Cosmological paradox in the dark of the night." *Science News*: February 23, 1991, p. 125.
29. "How to Go Back in Time." Michael D. Lomonick, *Time:* 13 May 1991, p. 74.
30. "Quasars: The Brightest and the Farthest." R. Cowen, *Science News*: vol. 139, May 4, 1991, p. 276.
31. "Maker of Worlds." David H. Freedman, *Discover:* July 1990, p.46.
32. "Cracks in the Cosmos." Ivars Peterson, Science News: vol. 139, p. 344.
33. "Galactic Chimneys." Andrew Chaikin, *Discover:* May 1991, p.24.
34. "Letters." *Science News*: July 27 1991, p. 51.
35. "Probing the Edge of the Universe." Sam Flamsteed, *Discover:* July 1991, p. 43.
36. "Going deep to fill in a blank piece of sky." *Science News*: vol. 139, p. 372.
37. Quasar quandary upsets status quo." R. Cowen, *Science News*: June 15, 1991, p. 375.
38. "Many Stars Are Born." *Newsweek:* 15 June 1992, p. 67.

39. "Virgo images suggest smaller universe." Ron Cowen, *Science News*: June 15, 1991, p. 381.
40. "ROSAT Revelations." Ron Cowen, *Science News*: June 29, 1991, p. 408.
41. *Science News*: April 21, 1990.
42. Sharon Begley with Daniel Glick, *Newsweek:* June 3, 1991, p. 47.
43. Ibid. 7.
44. "NASA discovery unlocks 'Big Bang' mystery." *Bangor Daily News:* 24 April 1992.
45. "'Big Bang' finding re-ignites religion, science debate." *David Briggs, Associated Press: Bangor* Daily News: April 25, 1992.
46. "The cosmic connection." *Popular Science:* April, 1991, p. 92.
47. "Bang! A Big Theory May Be Shot." Michael D. Lemonick, *Time:* January 14, 1991, p. 63.
48. "State of the Universe: If not with a Big Bang, then what?" Ivars Peterson, *Science News*.: April 13, 1991, p. 232
49. "Surveyor of the Universe." Andrew Chaikin, *Air and Space:* August/September 1991, p. 84.
50. "Big Bang Under Fire." Michael D. Lomonick, *Time:* September 2, 1991, p. 62.
51. "Squeezing Light for Precision, Speed." I. Peterson, *Science News:* May 30, 1992, p. 356.
52. "Cosmology Theory Compromised." Robert L. Oldershaw, Amherst College, Scientific Correspondence: *Nature*: 30 August 1990, p. 800.
53. "Galaxies – then you saw them, now you don't." Richard S. Ellis and Carlos S. Frenk, *Nature*: 30 August 1990, p. 790.
54. "Milky Way Monster." *Time*: July 27, 1992, p. 27.
55. "Exploring the Extreme Ultraviolet." Ron Cowen, *Science News*: May 23, 1992, p. 344.
56. "Looking For Lumps." Ron Cowen, *Science News*: May 8, 1993, p. 296.
57. *"Six Easy Pieces"* Richard P. Feynman, p. 136, Addison-Wesley Publishing.
58. Special Edition 1991. *Scientific American*: p. 6.
59. "Unified Theories of Elementary Particle Interaction." Steven Weinberg, Ibid., p. 32.
60. "The Number of Families of Matter." Gary J. Feldman and Jack Steinberger, Ibid. p. 32.
61. "On the Generalized Theory of Gravitation." Albert Einstein, Ibid. p. 40.

62. "The Inflationary Universe." Alan H. Guth and Paul Steinhardt, Ibid. p. 48.
63. "Particle Accelerators Test Cosmological Theory." David N. Schramm and Gary Steigman, Ibid. p. 62.
64. "The Mystery of the Cosmological Constant." Larry Abbott, Ibid. p. 72.
65. "Troubled Hubble still making discoveries." *Bangor Daily News*: 10 June.
66. "The Universe may be younger than thought." Ibid. 20 January 1985.
67. "A new way of making spectral redshifts." D.E. Thomsen, *Science News*: July 11, 1987, p. 22.
68. "A River Runs Through It?" Ron Cowen, *Science News*: December 12, 1992, p. 408.
69. "The Big Bang Never Happened." Eric J. Lerner, *Discover*: June 1988, p. 70.
70. "Particle Hunters." David H. Freedman, *Discover*: December 1992, p. 90.
71. Ibid. p. 92.
72. Ibid. p. 94.
73. Ibid. p. 98.
74. "Cosmological Controversy: Inflation, Texture, and Waves." Ron Cowen, Ibid. p. 328.
75. "A closer view of our galaxy's center." I bid, p. 334.
76. Hubble observations back merger theory." D. Pendick, *Science News*: June 5, 1993, p. 358.
77. Special Edition 1994, Scientific American.
78. "Tracking the evolution of galaxies." Ron Cowen, *Science News*: January 2, 1993, p. 15.
79. "Did Geminga Create Our Hole in Place?" R. Cowen, Ibid. p. 4.
80. "Science News of the Year." *Science News*: December 19 and 26, 1992, p. 432.
81. Ibid. p. 433.
82. "Nearby galaxy sheds light on dark matter." R. Cowen, *Science News*: June 12, 1993, p. 374.
83. Three's Company." Ron Cowen, *Science News*: August 28, 1993, p. 136.
84. "Dark matter: MACHO's in Milky Way's halo?" R. Cowen, *Science News*: September 25, 1993.
85. "Evidence of disks in middleweight stars." R. Cowen, *Science News:* 27 November, 1993.

86. "Chaos in Spacetime." Ivars Peterson, *Science News*: December 4, 1993, p. 376.
87. "New Challenger to the Big Bang?" Ron Cowen, *Science News*: October 9, 1993, p. 236.
88. "Bright Fires Around Us." Tim Folger, *Discover*: August 1993, p.18.
89. "Dark Matter: A cosmos that runs hot and cold." R. Cowen, *Science News*: P. 69.
90. "Quasar count poses dark-matter puzzle." Ron Cowen, *Science News*: June 19, 1993.
91. "The End of Physics." Science News Books, *Science News*: July 17, 1993.
92. "The Ultimate Vanishing." Tim Folger, *Discover*: October 1993, p. 98.
93. "Hubble cost to top $6 billion by 2005" Ann LoLorado, Baltimore Sun, *Bangor Daily News*.
94. "Strings and Mirrors." Ivars Peterson, *Science News*: February 27, 1993, p. 136.
95. "The End of Physics." Science News Books, *Science News*.
96. "Questioning a galactic star-forming model." R. Cowen, *Science News*: November 13,1993, p. 311.
97. "Starbirth model fixes our galaxy's age." R. Cowen, *Science News*: vol. 143.
98. "Finding Riddles of Physical Uncertainty." I Peterson, *Science News*: September 18, 1993, p. 180.
99. *Science News*: July 3, 1993.
100. "Finding the origins of the x-ray sky." R. Cowen, *Science News*: September 18, 1993. P. 183.
101. "Heavy elements found in interstellar gas." R. Cowen, *Science News:* p. 244.
102. *The Nature of Space and Time.* 63716 Library of Science, Newbridge Book Club, PO Box 6014, Delran, NJ 08075-9649
103. "Shattering the SSC vision: What next?" I Peterson, *Science News*: October 30, 1993.
104. "Cosmic dust can ferry in organic molecules." R. Lipkin, Ibid. p.278.
105. "Chemical pathway links stars meteorites." R. Cowen, *Science News*: November 6, 1993.
106. *Dark Matter Missing Planets and New Comets.* Ton Van Flandern, North Atlantic Books: p. 92.
107. Ibid. p. 370.

108. "Giant telescope aimed at cosmos." Matt Mygatt, *Bangor Daily News*.:20 August 1993.
109. "Dark darker than space found on galaxy edges." John Noble Wilford, New York Times, *Bangor Daily News*: 21 September 1993.
110. "Twinkles in the Dark." Michael D. Lemonick, *Time*: 4 October 1993, p. 77.
111. "Theory explains dark matter." John Noble Wilford, New York Times, *Bangor Daily News*: 11 October 1993.
112. "Partical physics: Stanford wins a B Factory." I. Peterson, *Science News*: October 16, 1993.
113. "Going for Glitz," Ivars Peterson, *Science News*: October 9, 1993. P. 232.
114. *The End of Physics*. David Lindley, Basic Books, A Division of Harper Collins Publishers, Inc.
115. "The Case of the Missing Neutron Stars." Tim Folger, *Discover:* December 1993, p. 36.
116. "The Road to Star Death." *Discover*: July 1993.
117. "Maps sharpen view of cosmic radiation." R. Cowen, *Science News*: January 29, 1994, p. 69.
118. *Tracking the evolution of galaxies*. Ron Cowen, Ibid. p. 77.
119. "Simply Plasma." Ivars Peterson, *Science News*: January 8, 1994, p. 30.
120. *Cold Fusion: the scientific fiasco of the century*. John R. Huizenga, Oxford University Press. Credited in Appendix I.
121. "Sunspot cycles." Clair Wood, *Bangor Daily News*: 14 January 1994.
122. "Private funds give SETI second life." *Bangor Daily News*: 14 January 1994.
123. *Quarks, Critters, and Chaos*. Jo Ann Shroyer, Prentiss Hall General Reference.
124. META Research Bulletin, vol. 1, no. 2, p. 20.
125. Ibid. vol. 5, #3
126. Ibid. vol. 5, #4, p. 64.
127. *The End of Physics*. David Lindley, Basic Books, A Division of Harper Collins Publishers, P. 3
128. Ibid. p.5.
129. Ibid. p. 11.
130. Ibid. p.131.
131. Ibid. p. 173.
132. Ibid. p.178.

133."New Eyes on the Universe." Bradford A. Smith, *National Geographic*: January 1994, p. 4.

134."Closing in on the Hubble Constant." R. Cowen, *Science News*: January 5, 1991, p. 4.

135."*The End of Physics.*" David Lindley, p.181.

136."*Reversing Diabetes*," Julian M. Whitaker, MD Warner Books.

137.*The Seven Mysteries of Life.* Guy Murchie, -- Houghton Mifflin Co., p. 315.

138.Ibid. p. 610.

139."Satellite retired after exploring Big Bang beginnings." Harry F. Rosenthal, *Bangor Daily News*: 24 December 1993.

140."COBE Sows Cosmological Confusion." *Science*: 3 July 1992, p.28.

141."Dim Stars Everywhere." Robert Naeye, *Discover*: May 1994, p.30.

142."Hubble finds massive black hole." Kathy Sawyer, Washington Post, *Bangor Daily News*: 26 May 1994.

143."Keck Telescope looks at the Big Bang." *Science News*: May 28, 1994, p. 349.

144."Star's eruptions reported." Paul Recer, *Bangor Daily News*: 1 July 1994.

145."Star change may revise theory on black holes." Paul Recer, *Bangor Daily News*: 31 May 1994.

146."Gathering String." Madhusree Mukerjee, *Scientific American*: June 1994, p. 16.

147."Studies suggest galaxies formed very early." R. Cowen, *Science News*: May 14, 1994, p. 311.

148."Sanity Check." Corey S. Powell, *Scientific American*: June 1994, p. 22.

149."Astronomers find heaviest elements detected in space study." *Bangor Daily News*: 8 July 1994.

150."Mysterious Rings Surround Supernova." R. Cowen, *Science News*: June 4, 1994.

151."Gotcha!" J. Madeleine Nash, *Time*: 9 May 1994, p. 69.

152."Waves from a Parallel Universe." Ivars Peterson, *Science News*: May 7, 1994, p. 300.

153."ASCA probes the x-ray universe near and far." R. Cowen, *Science News*: April 30, 1994, p. 277.

154."A Fresh Look at a Familiar Supernova." Ron Cowen, *Science News*: August 6, 1994, p. 90.

155."Ida's moon: A sharper view." Ibid.

156. "Astronomers say new evidence favors Big Bang Theory." John Noble Wilford, New York Times, *Maine Sunday Telegram*: 24 April 1994, p. 3c.

157. "The Keck Scopes Out the Legacy of the Big Bang." Ray Jayawardhana, *Science*: 15 April 1994, p. 346.

158. "Hole Meets Hole." *Discover*: July 1994, p. 18.

159. Tim Folger, *Discover*: July 1994, p. 24.

160. *Discover*: July 1994, p.86.

161. "Alternative Realities" William F. Allman, *U. S. News and World Report:* 9 May 1994, p. 59.

162. Michael D. Lemonick and J. Madeleine Nash, *Time*: 6 March 1995, p. 77.

163. "Searching for Cosmology's Holy Grail." Ron Cowen, *Science News*: October 8, 1994, p. 232.

164. "The quasar next door." *Science News*: October 8, 1994, p. 235.

165. "Heavenly chaos: A Star's erratic emissions." R. Cowen, *Science News*: February 18, 1995. P. 101.
"Making universes, constants our of nothing." I. Peterson, Ibid. p.102.

166. "How Stars Die: The Shocking Truth." Ron Cowen, *Science News*: February 18, 1995, p. 106.

167. "Gamma-ray bursts: The mystery deepens." AC Brooks, *Science News:* December 17, 1994, p. 404.

168. "After the Crash." Ron Cowen, Ibid. p. 412.

169. "Astronomical embarrassment of riches." Clair Wood, *Bangor Daily News*: 13 January 1995.

170. "Quasars without clothes." Ron Cowen, *Science News*: January 28, 1995, p. 56.

171. "Keck goes the distance for faraway galaxy." Ibid. January 14, 1995.

172. "Universe age may be revised." Washington Post, *Bangor Daily News*.

173. "Hubble eyes the Cartwheel." Ron Cowen, *Science News*: January 21, 1995.

174. "Dating the cosmos from super stars and relic radiation." *Science News:* October 22, 1994.

175. "Hubble offers clues to galactic evolution." R. Cowen, *Science News:* December 10, 1994.

176. Hubble finds dark matter still a mystery." Ibid. November 26, 1994.

177. "Searchlight on the Cosmos." Ibid. September 17, 1994, p. 188.

178. "Big Bang: Big Bust?" Alan L. Hausman, Letters, *Science News*: September 3, 1994.

179. "Roentgen's Universe." Cosmology, *Discover*: August 1994.

180. "Researchers use Hubble to trace noise from nearby galaxy to quasar." Doug Birch, Baltimore Sun, *Bangor Daily News:* September 24-25, 1994.

181. "Astronomers: The hills aren't as old as we thought." Robert Lee Hotz, Los Angeles Times, *Bangor Daily News*: 30 September 1994.

182. "Light halo hints at a galaxy's dark matter." R. Cowen, *Science News*: p. 119.

183. "Discovery produces Big Bang." David Chandler, Boston Globe, *Bangor Daily News*: 28-29 January 1995.

184. "Wave Packets." *Scientific American*: August 1993, p. 57.

185. "Brown dwarfs: Finding the lithium benchmark." R. Cowen, *Science News*. June 24, 1993, p. 389.

186. "Gravity's force: Chasing an elusive constant." I Peterson, *Science News*: April 29, 1995.

187. "Science's success also foremost enemy." Clair Wood, *Bangor Daily News*: 5 July 1996.

188. "Observatory spies some familiar compounds." R. Cowen, *Science News*: June 22, 1996, p. 389.

189. "Great wall not so great?" Ibid. p. 395.

190. "Deuterium provides a cosmic numbers game." R. Cowen, *Science News*: May 18, 1996, p. 309.

191. "A new instrument could spot faintest stars." R. Lipkin, *Science News*: May 11, 1996, p. 293.

192. "Age of the Cosmos: a first consensus." R. Cowen, Ibid. p. 292.

193. "Astronomers detect 14-billion-year-old galaxy." "Universe theory may need revision." *Bangor Daily News*.

194. "Galaxies become new stars." K.C. Cole, Los Angeles Times, *Bangor Daily News*.

195. "The Little Bang." Ron Cowen, *Science News*: June 24, 1995.

196. "Deep images favor expanding universe." R. Cowen, *Science News*: April 20, 1996, p. 246.

197. "Tracing the architecture of dark matter." R. Cowen, *Science News*: February 10, 1996, p. 87.

198. "The real meaning of 50 billion galaxies." Ron Cowen, *Science News*: February 3, 1996, p. 77.

199. "Whence the rays? Thence the rays." Astronomy, *Discover*: March 1996, p. 20.

200. "The answer in the voids." Jeffrey Winters, Ibid. p. 21.

201. "Explaining Everything." Madhusree Mukerjee, *Scientific American:* January 1996, p. 88.
202. "The vacant interstellar spaces." *Discover:* April 1996, p. 23.
203. "The Antimatter Mission." Gary Taubes, Ibid. p. 73.
204. "Detecting gas clouds in cosmic voids." "New spin on galaxy formation." "A star with jets?" Ron Cowen, *Science News:* July 1, 1995. P. 9.
205. "Further evidence of a youthful universe." R. Cowen, *Science News:* September 9, 1995, p. 166.
206. "Magnetic monopolies in matter." Ibid. p. 175.
207. "Strings and Webbs." Ivars Peterson, *Science News:* August 26, 1995, p. 140.
208. "Galileo encounters intense dust storm." Astronomy, *Science News:* September 16, 1995. P. 191.
209. "Opening the Door to the Early Cosmos." R. Cowen, *Science News:* August 3, 1996, p. 68.
210. "Eyeing Evidence of Primordial Helium." R. Cowen, *Science News:* June 17, 1995, p. 372.
211. "ASCA sheds light on galaxy formation." *Science News:* July 22, 1995, p. 54.
212. "Galaxy evolution: A multi-wavelength view." R. Cowen, *Science News:* June 10, 1995, p. 358.
213. "The Fingerprint of Creation." Donald Goldsmith, *Discover.* October 1992, p. 74.
214. "What Is Mater?" Erwin Schrodinger, Special Edition 1991, *Scientific American:* P. 16.
215. "Revisionists' view of solar system environs." R. Cowen, *Science News:* May 22, 1993, p.326.
216. "Milky Way Star Birth: Some far out action." R. Cowen, *Science News:* August 21, 1993.
217. "*Natural History of the Universe.*" Colin A. Ronan, Macmillan, p.38.
218. Ibid., p. 56, p. 78, p. 90.
219. "Shedding new light on a luminous galaxy." Ron Cowen, *Science News:* July 4, 1992.
220. "Cosmic Chemistry. Ron Cowen, *Science News:* November 22, 1996, p. 286.
221. "Alien Life, On Earth." Jim Wilson, *Popular Mechanics:* December 1996, p. 26.
222. "How to Light a Fire: Studies of the sun's corona heat up." Ron Cowen, *Science News:* August 31, 1996, p. 136.

223. "Glowing doughnuts flash high above storms." R. Monastersky, *Science News*: December 23 and 30, p. 421.
224. "In the Nursery of the Stars." Adam Frank, *Discover:* February 1996, p. 38.
225. "The Mediocre Universe." David H. Freedman, Ibid. p. 65.
226. "Weaving the Cosmic Web." Ron Cowen, *Science News*: September 23, 1995, p. 202.
227. "Crisis in the Cosmos." Sam Flamsteed, *Discover*: March 1995, p.66.
228. *"Feynman Lectures on Gravitation."* Richard P. Feynman, et al, Addison-Wesley Publishing Company, p. xv, xvi, and 6.
229. Ibid. p. 12.
230. Ibid. p. 19.
231. Ibid. p. 21.
232. Ibid. p. 22.
233. Ibid. p. 113.
234. Ibid. p. 149.
235. Ibid. p. 151
236. Ibid. p. 168.
237. Ibid. p. 173.
238. Ibid. p. 178.
239. Ibid. p. 187.
240. "Lone Stars." Ron Cowen, *Science News*: February 1, 1997.
241. "Fountain of Annihilation." Sharon Begley, *Newsweek*: May 12, 1997, p. 61.
242. "Does the Cosmos Have a Direction?" Ron Cowen, *Science News*: April 26, 1997, p. 252.
243. "From Soup to Us." Ron Cowen, *Science News*: June 7, 1997, p.354.
244. A Supernova Turns 10." Ron Cowen, *Science News*: February 22, 1997, p. 120.
245. "Books." *Science News*: November 15, 1997, p. 306.
246. "History of Physics." Lewis Pyenson, *Encyclopedia of Physics*: Second Edition, p. 505.
247. "Observational Cosmology Impacts Physics." Halton Arp, *Physics Essays*: vol. 8, no. 3, 1995, p. 350.
248. Ibid. p.351.
249. Ibid. p. 352.
250. Ibid. p. 353.
251. Ibid. p. 357.
252. Ibid. p. 360.
253. Ibid. p. 361.

254.Ibid. p. 364.
255."History of Physics." Lewis Pyenson, *Encyclopedia of Physics: Second Edition*, p. 519.
256.Ibid. p. 205.
257."Light from The Early Universe." Ron Cowen, *Science News:* February 7, 1998, p. 92.
258."New Budget Provides Lift for Science." J. Travis, Ibid. p. 87.
259."Reports raise questions about Martian rock." R. Cowen, *Science New:* January 24, 1998, p. 54.
260."Homing in on Milky Way's black hole." R.C., Ibid. p. 59.
261."Harpooning a Comet And Other New Space Probes." Tara Weingarten, *Newsweek:* December 15, 1997, p. 16.
262."The Universe, Ad Infinitum." John Allen Paulos, *New York Times:* 10 January 1998.
263."The Cosmos' Fate: World Without End." R. Cowen, *Science News:* January 3, 1998, p. 4.
264."Astronomers Aglow About Infrared Maps." R. Cowen, *Science News:* January 10, 1998, p. 20.
265."Scientist says tiny galaxy defying Milky Way's Pull." *Bangor Daily News:* February 16, 1998.
266.*"Times' Arrow and Archimedes' Point"* Huw Price, Oxford University Press, p. 83.
267."Filling Space" Martin, 3 March, 1999.
268."Entropy." Laszlo Tisca, *Encyclopedia of Physics:* Second Edition, p. 353.
269.*Bangor Daily News:* February 27, 1998, p. 1.
270.*META Research Bulletin:* Vol. 6, no. 4.
271."Circles in the Sky." Ivars Peterson, *Science News:* February 21, 1998, p. 123.
272."Infrared Map Helps Astronomers Measure Cosmic Energy." John Noble Wilford, *New York Times:* January 10, 1998.
273."Hipparcos finds hint of star streams." R. Cowen, *Science News:* February 28, 1998, p. 132.
274."Cosmologists in Flatland." Ron Cowen, Ibid. p. 139.
275.Letter to author, 30 June 1997, Halton Arp.
276.*Information Please Almanac 1998.* Houghton Miffin, p. 542.
277.Max Born's *Einstein's Theory of Relativity*, revised ed. Dover Publications, New York, 1962, p. 369.
278."Giant sky survey about to begin." *Bangor Daily News:* 9 June 1998, p. A7.
279."Weighing the Universe." Michael D. Lemonick, *Time:* June 15, 1998, p. 66.

280. "Ambitious sky survey gets underway." R. Cowen, *Science News*: June 13, 1998, p. 375.

281. "Modeling the whole universe." R.C., *Science News*: July 4, 1998, p. 11.

282. "Beyond Physics." W. Wayt Gibbs, *Scientific American*: August 1998, p. 20.

283. "Bright Lights, Big Mystery." George Musser, Ibid. p. 24.

284. "Starquake." Karen Southwell, *New Scientist*: 15 August 1998, p.26.

285. "If the force is with them..." Charles Seife, *New Scientist*: 12 September 1998, p. 4.

286. "Gravitation." John Friedman, *Encyclopedia of Physics*: Second Edition. P. 453.

287. "Another slinky candidate for galaxy seeds." P. Weiss, *Science News*: October 3, 1998, p. 215.

288. "Probing the heart of extragalactic jets." R. Cowen, Ibid. p. 214

289. "Infrared Camera Goes the Distance." R. Cowan, *Science News*: October 10, 1998, p. 228.

290. "A New, distant galaxy." R.C., *Science News*: November 7, 1998, p. 296.

291. "The greatest Story Ever Told. Is Cosmology Solved?" Ron Cowen, *Science News*: December 19 and 26, 1998, p. 392.

292. "Tiny galaxies have hearts of darkness." R. Cowen, *Science News*: January 16, 1999, p. 38.

293. "Surveying Spacetime with Supernovae." *Scientific American*: January 1999, p. 46.

294. "I Is The Law." Robert Matthews, *New Scientist*: 30 January 1999, p. 24.

295. "Escape from Earth" Charse Seife, *New Scientist*: 6 February, 1999, p.6.

296. "Shadowlands" Hazel Muir, *New Scientist:* 13 February, 1999, p. 5.

297. "Hit or Myth?" Harry Collins, Ibid. 12 September, 1998, p.37.